ORIGIN OF LIFE

Charles D. Schaper, Ph.D.

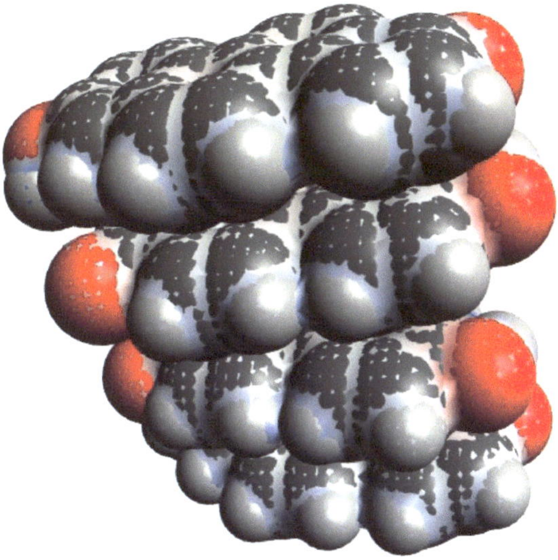

Molecular Sciences Publishing House

=======

Version 1.0:	July 26, 2020
Version 1.1:	September 13, 2020
Version 1.2:	October 11, 2020
Version 1.3:	November 5, 2020

Schaper, Charles
Origin of Life
Total: 233 total pages, 160 figures and 12 tables
Includes encoding, coding, decoding, and translation figures and tables
ISBN 978-1-7353721-6-7
I. Book design II. Title and Contents

Union City, CA, 94587
United States of America
cschaper@transferdevices.com

Contents

Biographical Sketch

Relevant to this subject matter, Charles Schaper's background includes training in engineering, chemistry, information theory, and nanotechnology, as follows:

- B.S. and M.S degrees in Chemical Engineering from the University of Connecticut, including a master's thesis with Professor Cooper on state-space design of control systems.

- Ph.D. in Chemical Engineering from the University of California, Santa Barbara with co-advisors Professors Seborg and Mellichamp, including a dissertation on stochastic robust control theory.

- Post-Doctorate research at the Information Systems Laboratory, Electrical Engineering Department, Stanford University working with Professors Kailath, Boyd, Franklin and Saraswat on applications of communications, computation, control, and signal processing to silicon semiconductor manufacturing.

- At Stanford University spanning fifteen years as academic research staff, electrical engineering, ultimately as senior research scientist working with Professors Kailath and Pease, which included a patented invention for the replication and transfer of microstructures and nanostructures.

- Founder and president of a nanotechnology start-up company on molecular transfer lithography, which included working as Principal Investigator on National Science Foundation grants.

- Admitted to medical school, M.D. program, for the 2019/2020 academic class at St. James Medical School, although have not started coursework yet, just external preliminary classes, with plans to defer the official start for another year.

Preprints

The path for the results described herein is documented by the following set of preprints, which is listed in reverse chronological order. It began from a mathematical investigation of a fever in response to exercise after vaccination, which was conducted as a personal curiosity, as are all the subsequent results. After about a six month break from the first two preprints, I investigated the underlying chemistry of fever first at the systems level, then the cellular level, and finally the nuclear level, which led to the results described herein on DNA structure and function.

1. Schaper, Charles (2020): Direct Binding of Complementary, Symmetric Ligands to DNA Nucleotide Pairs by Means of Transcription Factors. Preprints.org. https://doi.org/10.20944/preprints202009.0300.v1.

2. Schaper, Charles (2020): Intermolecular Enzymatic Encoding of Nucleic Acid, Steroid Complexes: A New Theory on the Chemical Origin of Life Based on Evidence of Structural Symmetry. https://arxiv.org/abs/2006.06429.

3. Schaper, Charles (2020): Intermolecular Binding of Functional Cyclic Compounds to RNA Nucleotides by an Ionic Process. Preprints.org. https://doi.org/10.20944/preprints202004.0522.v1

4. Schaper, Charles (2020): New Code for DNA Nucleotide Sequences. ChemRxiv. https://doi.org/10.26434/chemrxiv.12003786.v1

5. Schaper, Charles (2020): Structural Symmetry of DNA Nucleotides and Steroid Hormones. ChemRxiv. https://doi.org/10.26434/chemrxiv.11991567.v1

6. Schaper, Charles (2020): Endogenous Binding of Steroid Molecules to DNA Nucleotides by a Ca2+/PO4- Process to Enable Gene Transcription. ChemRxiv. https://doi.org/10.26434/chemrxiv.11868261.v2

7. Schaper, Charles (2019): Competitive Inhibition of Cortisol by Prostaglandins at the Ligand Binding Domain of Glucocorticoid Receptors. BioRxiv. https://www.biorxiv.org/content/10.1101/851501v7

8. Schaper, Charles (2019): Identifying Integrative Molecular Pathways for Predictive Modeling of Infectious Disease. BioRxiv. https://www.biorxiv.org/content/10.1101/835546v1

9. Schaper, Charles (2019): Analytic Model of fMRI BOLD Signals for Separable Metrics of Neural and Metabolic Activity. BioRxiv. https://www.biorxiv.org/content/10.1101/573006v1

10. Schaper, Charles (2019): Dynamics and Internal Control of Body Temperature in Response to Infectious Agents and other Causal Factors. BioRxiv. https://www.biorxiv.org/content/10.1101/566679v1

Research Monographs

Three research monographs have been prepared. The first monograph released describes the origin of DNA and the Genetic Code. The second monograph adds the origin of a cellular enclosure with genetic material. The third monograph, which is in pre-release, scheduled for January 2021 coordinates the results of the second monograph with new research results on systems-level regulatory function.

1. Schaper, Charles (2020): Origin of Life, ISBN: 978-1-7353721-6-7.

2. Schaper, Charles (2020): Design of DNA, Genetic Codes, and Life Function, ISBN: 978-1-7353721-0-5.

Provisional Patents

Charles Schaper has filed several personal provisional patents associated with this work, of which the application in approximate form was subsequently published on preprint servers. These provisional patents comprise the following set including one that covers the material described herein, submitted for review under confidential disclosure.

1. Charles Schaper, "Binding Steroid Molecules to DNA", Provisional Patent: US 62/977,216. Filed 3/9/2020.

2. Charles Schaper, "Binding Nucleic Acids", Provisional Patent: US 63/016,446. Filed 5/5/2020.

3. Charles Schaper, "Synthesis of Genetic Structures", Provisional Patent: US 63/027,320. Filed 5/19/2020.

4. Charles Schaper, "Design of DNA, Genetic Codes, and Life Function", Provisional Patent. Ref: US 63/045,105. Filed 6/28/2020.

5. Charles Schaper, "Origin, Design, and Synthesis of DNA, the Genetic Codes, the Biological Cell, and Life Forms", Provisional Patent: US 63/054,116. Filed 7/20/2020.

6. Charles Schaper, "Planar Steroid Assemblies for Life Origination, Genetic Design, and Biologic Materials", Provisional Patent: US 63/087,270. Filed 10/4/2020.

7. Charles Schaper, "Creating Encoded Genetic Information from Molecular Structure for Cellular Life Function", Provisional Patent: US 63/109,957. Filed 11/5/2020.

Preface to First Version

Presenting a comprehensive solution, the manuscript covers encoding, transmission, decoding and translation of DNA, including the identification of the origin of DNA and the mechanism of action. Coding and decoding tables are provided, as is the molecular basis for the structure and function of DNA. As described in the manuscript, DNA is co-synthesized with the steroid molecule in a co-enzymatic complex, and thus the work is also fundamental in describing the structure and function of steroid molecules relative to DNA molecules, as both are required to induce life function. The results are enabled from multiple discoveries that I have made and discuss herein. Furthermore, the construct is described of a digital framework that results from molecular structures and intermolecular coupling arrangements. Therefore, in meeting the challenges associated with the objectives of the contest, it should qualify as a winning solution.

I trust this solution to be accurate and correct as it is consistent with observable metrics, answering many detailed questions on DNA and its function. The results are described in the format of a first draft, and thus another edit would be helpful, and is underway, but all of the main results are presented in cohesive format permitting a new understanding of DNA, as well as its interaction with steroid molecules. Together, instantaneous life function is the result at the completion of the synthesis comprising the manufacture of a molecular structured communication path, enabling simultaneous structure and genetic function. The results are powerful, illuminating, and should impact along multiple levels of science, medicine, and philosophy.

Preface to Second Version

While the first version describes the design of DNA and the Genetic Code, as if that's not enough, the second version includes the design source of the bilayer cell membrane that encloses DNA and its complementary components, planar steroid molecules that enable life function. Moreover, as both DNA and its cellular enclosure are derived from the same assembly of starting material molecules, with the main difference being the binding pattern arrangement, the design methodology enables a single site solution to the origination of life. Furthermore, as of this writing, it appears that there are additional biologic materials that are derived from still other binding pattern configurations associated with the same assembly of starting molecules. Having a consistent design for DNA, the Genetic Code, the Cell and the molecular components to set all in motion, enhances the credibility of the overall solution.

In addition to the results on coordinating a molecular cell enclosure, DNA, and the genetic code, other updates to the second version from the first version include the methodology of genetic rearrangement strategies, such as replication, error correction, and splicing. The terminology has been updated as well in which reference to the starting material and its derivatives are termed "planar" steroid molecules to differentiate the structure from the modern-day usage of steroids and steroid hormones, which are non-planar. Further addressing this issue, a chapter is included on the evolution of molecular structures deployed in the transcription process to draw attention to the transition of planar steroid molecules to living cells by the loss of aromaticity and the gain of strategically positioned methyl groups.

Chapter 1

Introduction

The origin of life is inherent to central tenets of science and philosophy. Prospective theories have been developed and studied at great length [1, 2, 3]. To engage in this broad topic, as all cellular organisms of life contain DNA, many developments on the origin of life start with proposing mechanisms responsible for synthesizing DNA starting from basic materials and processes that would have come into existence about the time Earth was formed [4, 5]. In terms of hypotheses for the creation of the first DNA molecule, there are excellent articles written on the range of ideas, such as those that revolve around starting with RNA molecules and others that are based upon geological events [6, 7, 8, 9]. Thus, in putting forth a solution for the origin of both the structure and function of DNA, particularly as it relates to information content, as this exposition develops, it would be distractive to compare the new approach directly with all alternative developments, because of the wide range and varying levels of evidentiary basis [10, 11]. Therefore, in this introductory section, the objective is to characterize the constituents of a valid approach that reveals the design of DNA and the genetic code.

In developing a quantitative approach on the origin of life, a difficult problem to address immediately is the so-called "chicken and egg dilemma", as to which one came first [12]. For example, under consensus theories on transcription, DNA is required to form proteins, but proteins are needed by DNA to initiate transcription. Thus, there has been the speculation that DNA came about from an "RNA world", and this theory has taken the preeminent position in explaining the origin of life [13]. Here, this issue is resolved as the developments originate from a single molecule responsible for both structure and function.

A valid solution to the design of DNA should provide answers to detailed questions on the molecular structure of DNA. For example, detailed structural questions must be answered: why are purine and pyrimidine the bases that comprise DNA? How are the dimensions established? Why does adenine-thymine only have two internal hydrogen bonds, whereas cytosine-guanine have three internal hydrogen bonds? What is the molecular basis for the structure of DNA? What reading mechanism is responsible for accessing the information content of DNA? In the developments described in this manuscript, answers to such detailed questions are provided, including the molecular basis for DNA and its genetic code.

Most importantly and generally the largest stumbling block is the resolution of the source of the "code of life", that is the genetic code, the communication channel to regulate informational flow as encoding, transmission, and decoding. Literature acknowledges that this is the biggest problem for the current theories [14, 15, 16], and no other approach seems close to providing an answer. Here, I show a clear answer and comprehensive solution for the design of DNA and the genetic code, deriving a base four bit and communications framework for the transmission of encoded information in a molecular setting.

Ideally, a new approach to the design of DNA and the genetic code should have a solution for a cellular enclosure, that is the origin of the cell, which is often considered a second great challenge after the identification of the source for the genetic code. In this manuscript, I also present a solution for the origin of the cell that is consistent with the origin of DNA and the genetic code. From the same starting molecule and assembly, I show that one binding configuration leads to encoded DNA and the components for life function, while an alternative binding configuration leads to its cellular enclosure. Thus, the developments indicate that at the same place, at the same time, from the same material, there is an occurrence of DNA, the genetic code, the cellular membrane, and the complementary molecular constructs to enable function.

Chapter 2

Discovering a Molecular Primer of DNA

The encoding, transmission, decoding, and translation mechanisms for DNA were enabled by a discovery that I made at the initiation of the investigatory path. The fundamental result was initially described in my preprints, [17, 18, 19], after filing for provisional patents. These developments are critical to the developments in providing: (1) an experimental basis at the onset of the investigation; (2) the underlying structural and chemical interactions affiliated with the origination of DNA; (3) the symmetric and complementary associations that relate to the starting material for the design of DNA: that is, the molecular primer for DNA; (4) a reformulation of the concept of origin of life into a specific question on synthesis, which when answered enables an analytic framework from which to work, beginning with a complex for encoding molecular information systems.

To develop the new solution on the origin of DNA structure and function, the experimental evidence consistent with the resultant solution is presented along three lines, two of which constitute discoveries and the third is confirming evidence. The first comprises a structural analysis of DNA to illustrate that there is a common shape embedded within each of the four DNA dinucleotide pairings, which will serve as a beacon to develop the new solution. I show that DNA has embedded within it the shape of a four fused ring steroid molecule and the match is perfect along seventeen elements. Next, a functional analysis of DNA is performed, of which the discovery is that there is a perfect complementary intermolecular link to corticosteroids that provides the "missing" third hydrogen bond of the A-T base pairing, thereby matching that of the G-C base pairing. This match, in my opinion, is the most perfect complementary match in structural biology, as it comprises not only seventeen atoms but also alignment of functional groups. Thus, the interaction with steroid molecules relates to both structure and function. To consider that this match is not coincidental and is a real discovery, pharmacological efficacy data associated with the corticosteroids is analyzed to evaluate the predictive properties of the newly found intermolecular relationships.

2.1 Structural Symmetry of Steroid Molecules and DNA Nucleotides

In Figure 2.1(a), a front view of the DNA nucleotides is shown for each of the four base pairings as listed from top to bottom in the figure: Adenine-Thymine (A-T), Cytosine-Guanine (C-G), Thymine-Adenine (T-A), and Guanine-Cytosine (G-C). Also included in the illustration is the inter-nucleotide hydrogen bonding of the base pairings, of which there are three hydrogen bonds for the C-G and G-C pairs, but only two inter-nucleotide hydrogen bonds for the A-T and T-A base pairs. It is of importance that at the carbon two position of thymine, there exists a functional oxygen element, a ketone, but there exists no complementary element on adenine, for example a hydroxyl or amine group, that could offer a link as a hydrogen bond.

To describe the discovery of a four fused ring structure in the shape of a steroid molecule embedded in DNA, the result is presented as the red overlays shown in Figure 2.1(b) highlighting the match, which I find to be extraordinary. This overlay is obtained by connecting a total of seventeen oxygen, nitrogen, and carbon elements of the base pairs in a particular manner to produce a tetra-ring structure for each of the four DNA nucleotide pairs. This can be achieved for each base pair and the match is in perfect agreement with each of the four Watson-Crick dinucleotides. The four-ring molecular structure is exactly in the shape of a steroid molecule, of which this match was originally identified in the preprints [17, 18]. The orientation of the steroid molecule is also in perfect correspondence when analyzing the front for the A-T and G-C pairings, where a purine molecule starts the association on the 5' side, while on the back the orientation is the same for the T-A and C-G pairings when the pyrimidine molecule starts the 5' pair. For each, there is a lateral stretching of one of the rings where the purine and pyrimidine come into alignment through hydrogen bonding. Thus, this association is a candidate basis for DNA of which any approach that seeks to explain the origin of DNA must identify, and thus the concepts developed herein begin from the steroid molecule, as it possesses a perfect match and is thus a candidate for the structural basis of DNA.

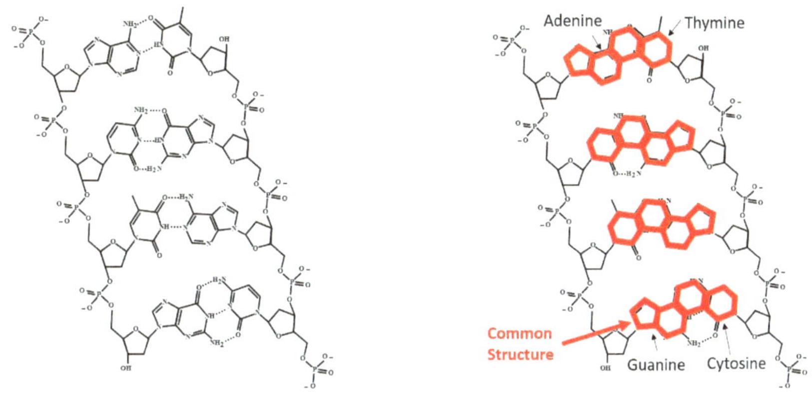

(a) ACTG (b) ACTG with overlay

Figure 2.1: (a) The four Watson-Crick DNA nucleotide pairs presented from top to bottom are Adenine-Thymine (A-T), Cytosine-Guanine (C-G), Thymine-Adenine (T-A), and Guanine-Cytosine (G-C). (b) By connecting seventeen elements of carbon, oxygen and nitrogen, there is a common structure in the shape of a tetra-ring steroid molecule within each nucleotide pair [17, 18].

2.2 Intermolecular Binding of Corticosteroids with Thymine

A significant consideration in this line of discoveries is identifying the reason for the A-T pairings possessing only two internal hydrogen bonds whereas the C-G pairings have three. Continuing with the relation between steroids of biological relevance and DNA, and even prior to the determination of the steroid molecule as a potential basis candidate for DNA, I noted that cortisol is perfectly complementary to the A-T pairings [17]. This observation was the kick-off point from which my interest in the relation between steroid molecules and DNA began. There is a mid-molecule hydroxyl group on the C-ring, as it is called, of cortisol which when aligned to A-T would form the missing third hydrogen bond with the available ketone group of thymine, ignoring for the moment the strategically positioned methyl groups that might get in the way. Moreover, the end groups of cortisol were also arranged in an interesting manner with respect to the sugar group and the phosphodiester chain. This relation of not only symmetry but also complementary match is illustrated in Figure 2.2. Also included is a reverse analysis on the C-G pairings, which already have three internal hydrogen bonds. If the complementary match between cortisol and the A-T dinucleotide is relevant, then it would stand to reason that there would be a biologic steroid hormone that matches in some manner to C-G. And, indeed there is, as shown in Figure 2.2, testosterone-like hormones, which do not have a hydroxyl group on the C-ring, align perfectly with C-G, since the end-groups are effective with the phosphodiester chain. Interestingly, like cortisol, testosterone also has two methyl groups in the same position as cortisol, which hinder a hydroxyl association through hydrogen binding with thymine. The orientation also is relevant as the complementary match is met on the front view when purines start the link to the 5' strand, but on the back with pyrimidines.

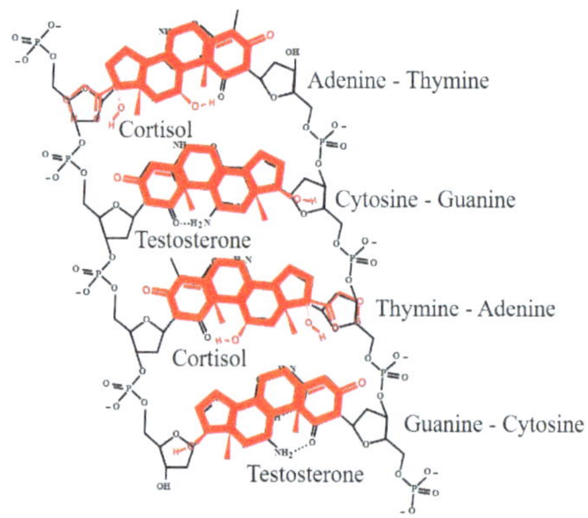

Figure 2.2: Consistent with the result of the shape of a common steroid molecule embedded within DNA nucleotide pairs, the intermolecular bonding capability is examined with respect to thymine and the steroid hormones. To make this study, Cortisol is representative of steroid molecules which have a functional hydroxyl group at a specific site on what is called the C-ring, and testosterone is representative of steroid molecules that do not have a hydroxyl group on the C-ring. Now, the objective is to overlay the steroid cortisol such that including the potential for intermolecular hydrogen bonding there would results three hydrogen bonds with the A-T and T-A nucleotide pairs. Since G-C and C-G already have three hydrogen bonds, an overlay of testosterone with enable intermolecular bonding and an equivalent set of three hydrogen bonds will match that of the cortisol A-T and T-A associations. Note that for both cortisol and testosterone, the end groups are available for connection with the phosphodiester backbone. In addition, it is noted that the same orientation of the methyl groups to the nucleotide is reached when cortisol forms the intermolecular bond with T-A on the opposite side as that corresponding to A-T. The same orientation is seen for testosterone and C-G versus G-C.

2.3 Correlation with Pharmacological Efficacy

Through the symmetric and complementary relation of the steroid hormones with DNA, a predictive capability results relating to pharmacological efficacy of synthetic corticosteroids, which indicates that the observation of a structural match to produce the third hydrogen bond with A-T has a functional purpose. In an experimental analysis, the intermolecular strength of the hydrogen bond with thymine is examined by comparative assessment of the potency of dexamethasone, prednisolone, prednisone relative to cortisol. The handbooks indicate that the relative activity of dexamethasone : prednisolone : cortisol is approximately 25 : 4 : 1 and that prednisone, in its native form, is inactive as it must be converted into prednisolone in the liver before it can be used [20]. By analyzing the strength of the potential hydrogen bond with thymine, the pharmacological efficacy data is predictable. In Figure 2.3(a), the structural relationship of prednisolone to thymine is indicated, in which it is shown that it is possible to form an intermolecular hydrogen bond with thymine. However, in Figure 2.3(b), the relationship of prednisone to thymine is indicated, in which it is not possible to form an intermolecular hydrogen bond. Note that there is just a single atom difference between prednisolone and prednisone, but it is at a critical point, the hydroxyl group that would bind onto thymine. This precise result is consistent with the efficacy of prednisone, which experimentally shows no activity until it is converted into prednisolone.

Further experimental data is consistent with the hypothesis that the intermolecular bond between steroids and the A-T dinucleotide is of a functional nature. In Figure 2.3(c), the relationship between the steroid hormone cortisol and thymine is indicated, which can form an intermolecular hydrogen bond, and with respect to prednisolone, the only difference is an extra double bond in the A-ring of prednisolone. The additional double bond will aid in stabilizing the intermolecular hydrogen bond of thymine and prednisolone and may also provide favorable structural arrangement in terms of the angle. Further, by examination of the chemical relation of dexamethasone in Figure 2.3(d), the hypothesis would indicate that activity is further enhanced due to the fluoride group stabilizing the intermolecular hydrogen bond even more so than prednisolone, thereby increasing its relative activity.

2.4 The Primer

My initial discovery both of a candidate for the structural basis of DNA as the steroid molecule, and the location for the missing third hydrogen bond of adenine-thymine as found in the corticosteroids, enables a guide for further discovery on the origins

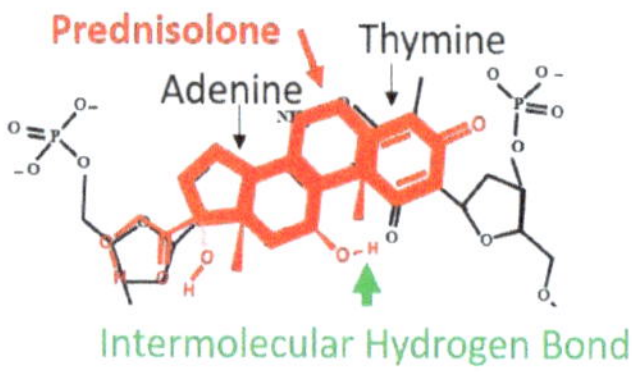

(a) Prednisolone

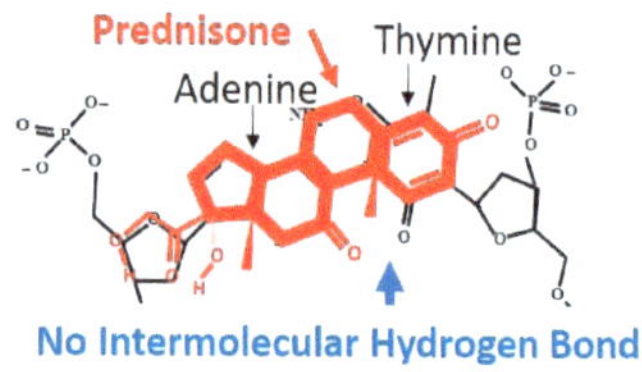

(b) Prednisone

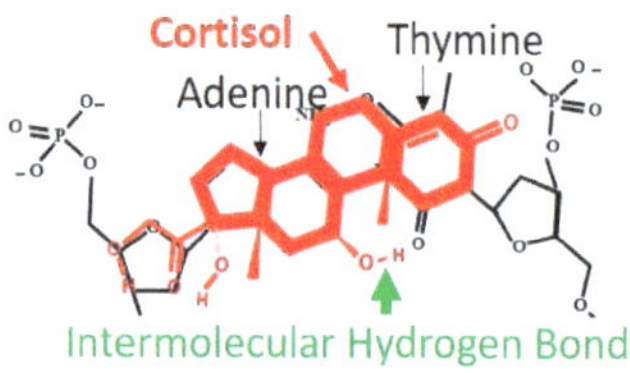

(c) Cortisol

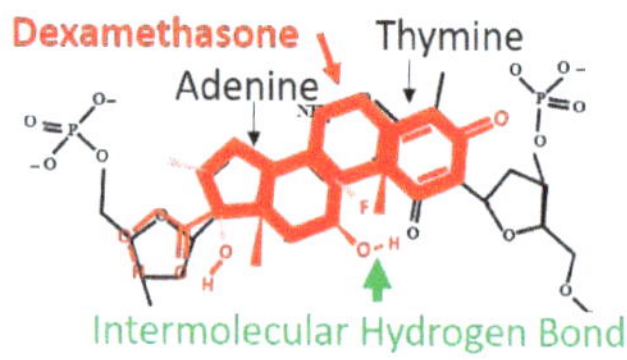

(d) Dexamethasone

Figure 2.3: To indicate that the structural and complementary relation between cortisol and the adenine-thymine pair is functional, an analysis of the impact on the pharmacological efficacy of the corticosteroids is assessed: (a) There is the capability to form an intermolecular hydrogen bond between prednisolone and thymine, and thus one would expect activity. (b) It is not possible to form an intermolecular hydrogen bond between prednisone and thymine. Indeed, prednisone is inactive and must be converted to prednisolone for prednisone to be effective. This result provides evidence in proving the claims of this manuscript. (c) The overlay of the nominal steroid hormone cortisol is shown capable to form an intermolecular hydrogen bond with thymine. With respect to prednisolone, the double bond in the ring adjacent to thymine aid in enhancing the intermolecular hydrogen bond strength, and thus increased potency is expected for prednisolone relative to cortisol, which is indeed the case in practice. (d) Dexamethasone can form an intermolecular hydrogen bond with thymine. In addition, the fluoride group and double bond will aid in the potency, and thus it would be expected to have greater potency than prednisolone, and cortisol, which again aligns with the efficacy observed in practice

of DNA and its life function by reshaping a vague concept into a specific objective of synthesis, for example: Prove the relevancy of the symmetric, complementary match between cortisol and the A-T dinucleotide. To address that objective, it is necessary to go all the way back to the beginning, which in turn will resolve the problem on the origin of life. This reformulation can guide the synthesis, and encoding procedures, such that the relation between steroid molecule and DNA is never lost. Thus, the molecular primer used in the development of the synthesis strategy, and its accompanying encoding scheme are based on the following set of hypotheses: (1) the steroid molecule is the basis for the structure of DNA as purine and pyrimidine will form three of the four rings, and its integration through hydrogen bonding, will form the fourth ring; (2) the planar steroid molecule and the DNA molecule are coupled, and were always coupled, capable of coordinated actions by alignment of the ring structures; (3) intermolecular hydrogen bonding between a C-ring hydroxyl group of cortisol-like steroids with the ketone group of thymine must always be maintained throughout the synthesis strategy so as to establish functionality in the end result. Thus, communication between the steroid molecule and the DNA molecule, which have a common shape, is established through a structural organization, that is the four-ring system, and a chemical organization, that is the intermolecular hydrogen bond. In the following chapters, which describe the origin of DNA and the genetic code, and which enable an origin for life as the developments uncover a source for the cell membrane, these connections on steroid-DNA start from the beginning, and are never lost.

Chapter 3

Molecular Complex for Encoding DNA

Shown for the first time, a rich array of assemblies and binding patterns of a planar steroid molecule results in several relevant biological constructs for life origination, including DNA, the genetic code, and the phospholipid bilayer cellular enclosure. In this chapter, I will focus on the arrangement of an assembly of planar steroid molecules and a binding pattern that results in DNA and its base four genetic code. In addition, this particular design also results in a set of complementary planar steroid molecules that are co-synthesized with DNA, which induce life function operations such as transcription and translation from the point at which the DNA double helix is formed. Other binding patterns and assemblies, and their biological outcome after synthesis, which are relevant to the viability of DNA, will be discussed in later chapters.

3.1 The Starting Material

The starting material for the creation of DNA and the genetic code is the planar steroid molecule, $C_{17}H_{10}O_2$, shown in Figure 3.1(a). The molecule is comprised of four fused rings, with a hydroxyl group positioned on the six-carbon ring at one end, and a ketone group on the five-carbon ring at the other end. The side view of the starting material is indicated in Figure 3.1(b) to show a planar structure. Appendix A.1.4 discusses sources for this starting molecule, such as chrysene.

The ring structure of the planar steroid molecule is sufficient to develop a base 4 code. In Figure 3.2(a), the four possible orientations of the four ring structure are presented. In Figure 3.2(b), a color coded representation, which will be used throughout the derivations, is presented for each of the configurations. Each of the rectangles represents a ring and its orientation. For example, orange represents the

(a) Starting Material (b) Side View

Figure 3.1: (a) The planar steroid molecule, $C_{17}H_{10}O_2$, is the starting material responsible for the origination of DNA, the genetic code, the cell membrane, and other biologic materials. Note that the ends of this planar steroid molecule are complementary to each other in terms of hydrogen bonding; (b) the side view of the molecule indicates planarity and the functional groups are oriented in the same plane as the rings.

five-carbon ring at the end; blue represents the six-carbon ring at the other end; cyan represents the six-carbon ring which is located towards the back when viewed from the frontside; green represents the six-carbon ring located forward when viewed from the frontside. The orientation of the end molecules, whether forward or back, can be determined from the orientation of the inner rings of the molecule, as to the sequence of four rings as back-forward-back-forward, or forward-back-forward-back. For example, the upper molecule in Figure 3.2(a) is forward-back-forward-back, and since the left-most ring is the five-carbon ring, the coloring is orange-cyan-green-orange, which is one of the arrangements in the coloring key of Figure 3.2(b).

3.2 Paired Ring Structures

To formulate the base four code, the ring structures will now be paired together as a dimer. Hydrogen bonding secures the dimer on the ends when oriented as shown in Figure 3.3(a). The alcohol group on the six-carbon ring interacts with the ketone of the five-carbon ring. The two internal rings of each dimer molecule are aligned.

For the construction of the dimer structure, which will be used to create DNA, the planar steroid of Figure 3.1(a) works for several reasons: (1) it is a planar molecule and can be stacked, self-assembled, one on top of each other; (2) it has a hydrophobic interior and hydrophilic ends, and thus a unique environment is maintained within the interior of the molecule while the ends stabilize within a water environment; (3) the molecule has significant chemical activity with multiple double bonds; (3) It can form a dimer as indicated in Figure 3.3(a), of which the dimers can then be stacked,

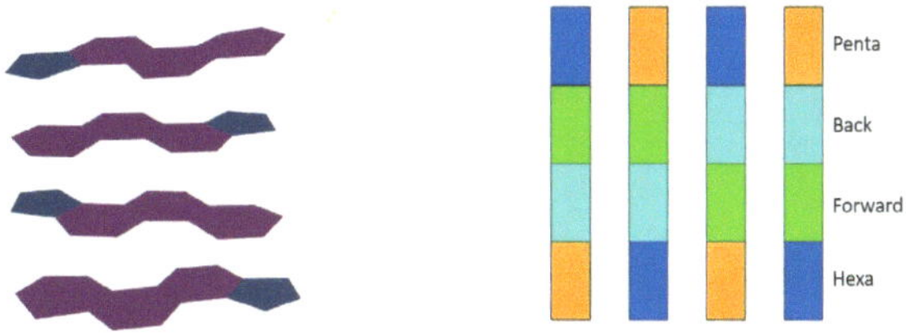

(a) Orientations (b) Encoding Key

Figure 3.2: (a) The representation of the ring structure comprising the starting molecule, $C_{17}H_{10}O_2$. It has four orientations from the viewers perspective; (b) A graphical illustration of the four configurations of the ring structure from a fixed reference. The color coding on the rings consist of the five-carbon (denoted as "penta") ring on one end and the six-carbon (denoted as "hexa") ring on the other end; and in the middle rings, the six-carbon rings are either forward towards the viewer, or back, away from the viewer. This orientation can be seen from the ring representation. The encoding key is developed in correspondence with the four orientations of the rings shown in (a). Each boxed rectangle represents a ring, and there is a one-to-one map of the ring representation of (a) and the encoding key of (b).

as indicated in Figure 3.3(b), thus forming long chains of repeating units; (4) the rings can be oriented such that the inner two rings are aligned and symmetric as indicated in Figure 3.3(c); (4) it has the same seventeen element match of the primer structure of DNA; (5) the dimer representation of the starting material has four and only four interdimeric configurations, and thus can produce a base four genetic code. The corresponding color coded key for the ring structure is shown in Figure 3.3(d). Note that the colors are opposite on the ends and matched in the middle.

In this paired configuration, there are four orientations possible of the same structure, when viewed from the same perspective, for example left to right. This is illustrated in Figure 3.4, which shows at the left, the sequence top to bottom as, blue, green, cyan, orange in 3.4(a) which according to the represents the penta-ring that has a ketone group, forward six-carbon ring, back six-carbon ring, hexa-ring, which has an alcohol group. This configuration can then be rotated about the horizontal and vertical axis to arrive at the other three configurations. These configurations will be used in creating a unified structure not only for DNA but also complementary planar steroid molecules, one of which is the same as the starting molecule.

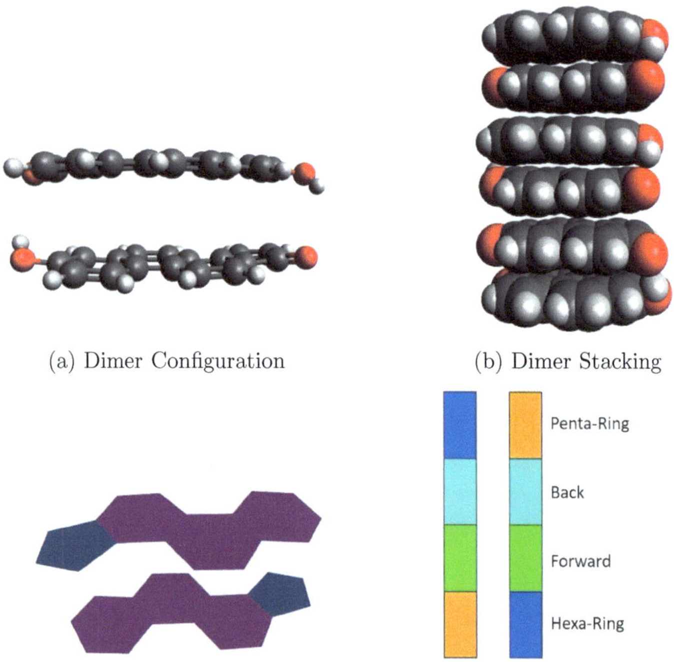

(a) Dimer Configuration (b) Dimer Stacking

(c) Ring Structure (d) Color Key

Figure 3.3: (a) The dimer configuration of the starting material. Note that it is functionalized in a complementary way at each end, one by a hydroxyl group and the other by a ketone, thus hydrogen bonding secures the two molecules; (b) stacking of three dimers represented with van der Waals spheres; (c) the ring structure orientation of the dimers; (d) the color key coding for the structures, which is used in the coding analysis. This coding key of using rectangles to represent a ring structure is used throughout this manuscript.

3.3 Sequence of Pairs

To build-out the assembly, the paired dimer molecules of Figure 3.4 will be paired to each other. Thus, for example, the sequence will be established as 3.4(b) followed by 3.4(c) followed by 3.4(a) followed by 3.4(d) and so forth. Before discussing the sequence, I will describe the chemistry of the pairing.

As indicated in Figure A.6, the coordination of the paired building blocks enables a sequence of effective coupled reactive zones, comprising a confined environment that

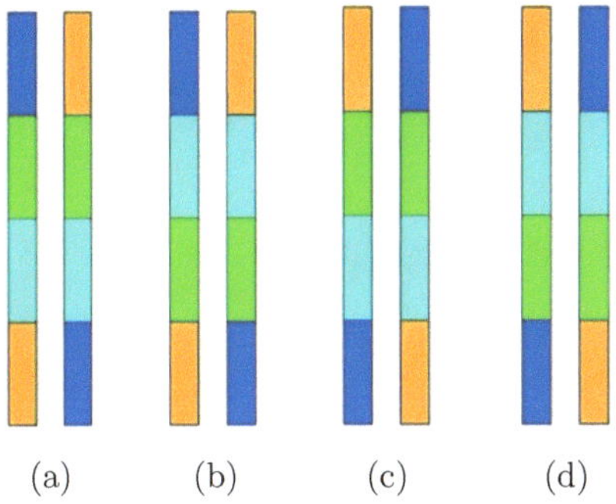

Figure 3.4: The color key for the four orientations of the dimer relative to a reference point. In (a) to (b), the dimer is rotated about both the horizontal and vertical axis; (a) to (c) is rotated about its vertical axis; (a) to (d) is rotated about the horizontal axis. The relationship between the opposing faces of the dimer will be used to determine the nature of the intermolecular link.

enables the potential for enzymatic capability in which the temporary formation of bonding arrangements to permit intermolecular structures that can establish element substitution, the degradation of chemical bonds to reduce intramolecular stress, and the differentiation of the templates into their individual capacity as a nucleotide base, preventing amination through hydrogen bonding, for example. These developments regarding the formation of DNA through co-enzymatic complexes within enclosed environments are discussed in Appendix A.

For encoding DNA and the complementary planar steroids, the main concept is that in stacking the next pair of dimers on top of the initiating pair of dimers, the relative orientation will define ultimately the "letters" of the nucleotide pair, whether it is adenine-thymine, thymine-adenine, guanine-cytosine, or cytosine-guanine. There are exactly four configurations, as indicated in Figure 3.6, when placing the second set of molecules on top of the first set of molecules, and as will be seen later, one molecule of the dimer will define the nucleotide pair, and the other will define the complementary steroid molecule that will interact with the DNA dinucleotides to trigger replication and transcription. Thus, for the configurations of Figure 3.6, starting from the bottom, Molecule 2 will be paired with Molecule 3 during the synthesis steps. Depending upon a subsequent binding pattern, one molecule will result in a DNA dinucleotide, and the other molecule will result in its complementary planar steroid. For example, one will form A-T dinucleotide and the other will form a planar steroid that has a hydroxyl group positioned such that it can form the third

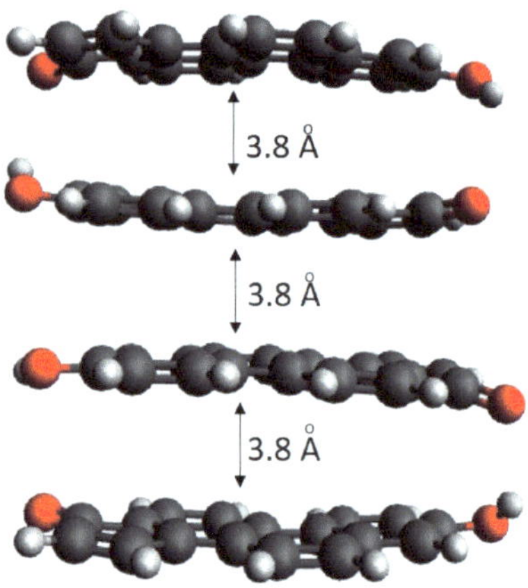

Figure 3.5: The stacking of two dimers is indicated to show a uniformity of distance, including between interdimeric interactions. The proximity of one molecule to another of approximately 3.8 Å enables the potential for influence in the synthesis reactions. Each atom along the periphery is accessible to the environment. The ends are more hydrophilic than the interior, which enables localized environments for synthesis.

hydrogen bond by coupling with thymine.

Since there are four orientations for each ring structure, the dimeric structures pairing to each other will result in 16 relative configurations. This is illustrated in Table 3.1, in which a letter code is used for each molecule: P denotes the five-carbon end ring; B denotes a back orientation of the inner six-carbon ring; F denotes a forward orientation of the six-carbon ring; K denotes the six-carbon end ring. The various orientations are obtained by rotating and flipping, and since there are four possibilities, it will be possible to obtain the next letter independent of the starting point of the initial dimer.

Remarkably, that is sufficient to create a encoded DNA molecule as well as its complementary planar steroid molecule, which will enable genetic replication and transcription. It is possible to create any set of letters from AGCT, as well as its planar steroid complement, by using the strategy of pairing and orientation, which is an amazingly simple process, as it needs to be! Moreover, in addition to the base 4 code associated with the orientation of the pairings, another code is defined, which is that of the planar steroid molecules that complement the A-T or C-G dinucleotides.

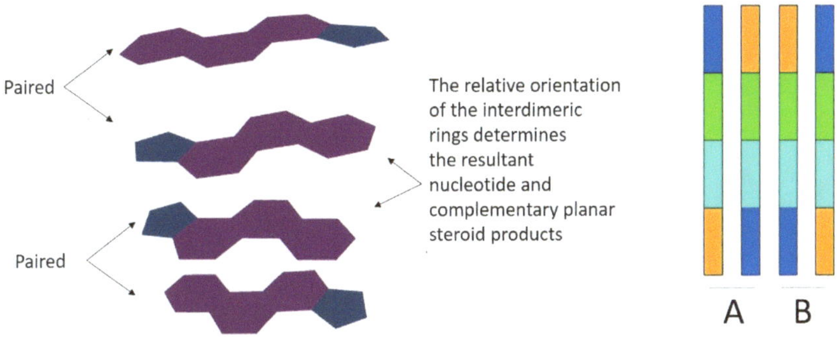

(a) Pair of dimers (b) Color key

Figure 3.6: (a) The resultant identification of the DNA nucleotide sequence and its complementary planar steroid products are determined by the relative orientation of the opposing faces of dimers formed during a self-assembled stacking process. (b) The color key associated with a pairing of dimers.

As will be developed, the planar steroid molecules are coordinated with amino acids during the transcription process, and thus will provide the specificity required for protein translation, as well as an inherent capability to induce strand separation.

3.4 Encoding through Ring Orientation

When one dimer is appended to an adjacent dimer, there is one of four orientations associated with the ring structure, as indicated in Figure 3.4. The orientation will define which one of the nucleotides from the set A-G-C-T results, and it will also determine the structure of the corresponding planar steroid. The orientation defines whether intermolecular binding between the opposing molecules is feasible due to the ring structure, as indicated in the proceeding chapter discussing the molecular primer of DNA - on reason why that chapter is important to the developments. If intermolecular hydrogen bonding is feasible, then the DNA dinucleotide will be A-T or T-A, depending upon whether the penta-ring or hexa-ring is first. The complementary planar steroid molecule will have a hydroxyl group on the C-ring. Otherwise, if intermolecular hydrogen bonding is not feasible, then the dinucleotide is G-C or C-G, depending upon whether the penta-ring or hexa-ring is first. The corresponding planar steroid molecule will be the same as the starting planar steroid material.

Reference	Next Molecule Options
PBFK	PBFK, PFBK, KFBP, KBFP
PFBK	PBFK, PFBK, KFBP, KBFP
KFBP	PBFK, PFBK, KFBP, KBFP
KBFP	PBFK, PFBK, KFBP, KBFP

Table 3.1: There are 16 possible configurations of the interdimeric opposing faces when pairing dimers. For each dimer configuration, there are four possible orientations of the opposing face of the next dimer. Since there are four possible configurations for each dimer, the total is thus four times four. However, when the orientation of one reference dimer is fixed there are only four possible options for the orientation of the opposing molecule of the next dimer.

For intermolecular hydrogen bonding to occur. the ring patterns can define whether it is feasible from the standpoint of the molecular primer, in which the C-ring couples into the A-ring. This can be expressed in color coded comparison of the 16 possible ring structure orientations as indicated in Figure 3.8, which indicates that of the 16 possible configurations, 8 will have hydrogen bonding capability, and the other 8 configurations will not have hydrogen bonding capability. There are sixteen configurations in dimeric representation, because each configuration of ring structures has four possible adjacent configurations.

In addition, for each configuration, there corresponds 2 configurations that will have intermolecular hydrogen bonding capability, and 2 configurations that will not have intermolecular hydrogen bonding capability. This is illustrated in Table 3.2. From the sequences, it can be seen that if the next molecule starts with the same molecule, then the correspondence has to perfectly matched in order for hydrogen bonding to occur; or if the next molecule is not the same, then it has to be completely out of phase in order for hydrogen bonding to occur. For example, if the ring structure of the outward facing molecule of a dimer is KFBP, then to have hydrogen bonding, the opposing face of the next dimer would have to be either the same KFBP, or completely opposite, PBFK. Thus the result of the dinucleotide associated with this orientation would be either A-T or T-A, which would then be determined based on the first ring: if starting with K, which is the six-carbon ring, the result is T-A; otherwise, if it starts with P, which is the five-carbon ring, the result is A-T.

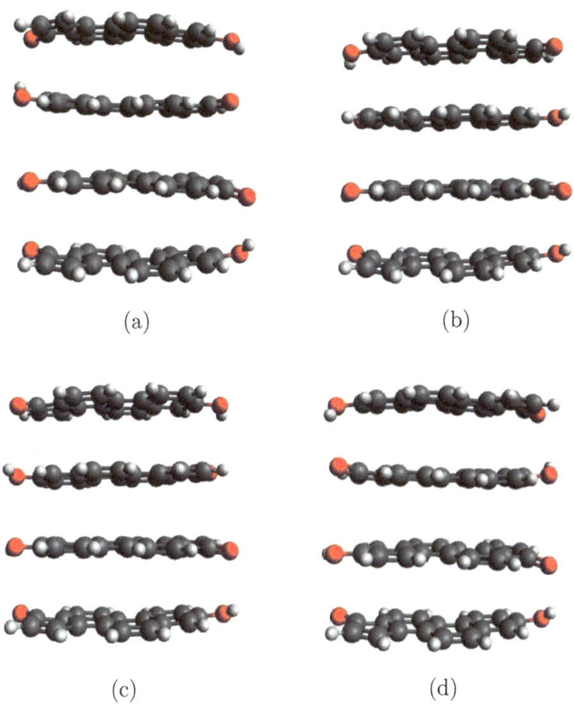

Figure 3.7: The orientation of the adjacent molecule of the second pair with respect to the molecule of the first pair will define the final result. There are four orientations possible. The coupling elements between the adjacent molecules, that is the nature of the intermolecular link, will be used to define the nucleotide product as well as the complementary planar steroid. Note that there are four rings that define each structure, and that there are four and only four dimer pairings. Thus starting with the upper dimer of (a), the difference with (b) is a rotation of the upper dimer with respect to the lower dimer; (b) to (c) involves a flipping of the upper dimer relative to the lower dimer; (c) to (d) involves a rotation of the upper dimer.

3.5 Summary on Encoding DNA

The molecule, the assembly and the binding pattern has been shown that result in a base 4 code. Moreover, as will be shown, the binding arrangement not only results in DNA, but also in a set of complementary planar steroid molecules that enable

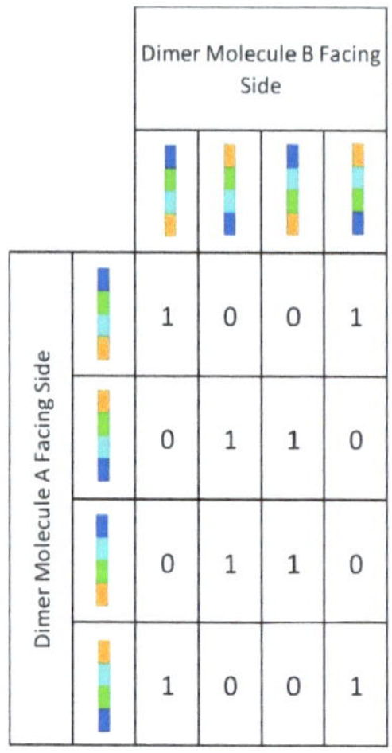

Figure 3.8: The intermolecular orientations of the ring structures of adjacent dimers that will lead to coupled hydrogen bonding. Out of the 16 total configurations, eight will lead to intermolecular hydrogen bonding. These couplings enable a base 4 code for the nucleotides, as well as two types of complementary planar steroid molecules.

life function activities, such as transcription, replication, and the genetic code. The encoding of a four-ring molecule which is capable of dimeric formation has been shown with alignment of the internal rings, and when configured as a stacked assembly with interdimeric association, in which the facing molecules of consecutive dimers have co-synthesized properties, result in an independent sequence. Other binding patterns and assemblies of the planar steroid are possible, and as will be shown result in useful biological materials, such as the cell, but are not encoded like that of DNA. Long sets of planar steroids can be strung together, and thus long sequences of DNA is feasible. It can also be represented as a long sequence of random dimer pairs of molecules, as indicated in Figure 3.9. This will result in a long DNA chain, and a set of complementary planar steroid molecules, which will thus enable structure and function, respectively. It will be seen that the planar steroid of Figure 3.1(a) not only act as the genesis for the creation of DNA, but also impart function into DNA, and bring it all to life.

Thus, by using this mechanism of assembled paired representations in specific binding configurations, long sequences can be arranged in which every other pair will have an orientation drawn from a set of four possibilities. For example, the

Referencing Pair	Next Pair Intermolecular Bonding	Next Pair No Intermolecular Bonding
PFBK	PFBK, KBFP	KFBP, PBFK
KFBP	PBFK, KFBP	KBFP, PFBK
PBFK	KFBP, PBFK	PFBK, KBFP
KBFP	KBFP, PFBK	KFBP, PBFK

Table 3.2: This table indicates the orientations that will lead to intermolecular hydrogen bonding when the pairs are stacked together, or no intermolecular hydrogen bonding of the complementary planar steroid to resultant nucleotide. The intermolecular bonding will define whether the resultant nucleotide pairing is A-T or C-G, and whether the planar steroid product is A-T complementary or C-G complementary. This code defines ring orientations as a design strategy to impart a base four digital code into a molecular structure.

sequence of paired molecules is indicated in 3.9 which for example has a first base 4 code set at molecules 1 - 3 - 5 - 7 ..., and a second base 4 code set at molecules 2 - 4 - 6 - 8 .... However, the first base 4 code set is correlated to the second base 4 code set because they were presented as a dimeric pair. Thus, the result will be one independent base 4 code set, and a second dependent set of planar steroids that can be derived from the first set. We will see that the first set corresponds to A - G - C -T DNA, and the second set will be the planar steroid molecules. Thus, pursuant to the discovery discussion of a molecular primer in Chapter 2, which notes the structural and complementary correlation of DNA nucleotides and steroid molecules, the planar steroid molecule is the genesis of DNA, and as will be seen later in this manuscript, also responsible for the cell membrane and other biological materials. Alternative binding patterns of the encoding complex are considered in Chapter 10. It is important to consider the ring structure of the planar steroid molecule as the exclusive defining characteristic that enables the base four code of DNA nucleotide sequences.

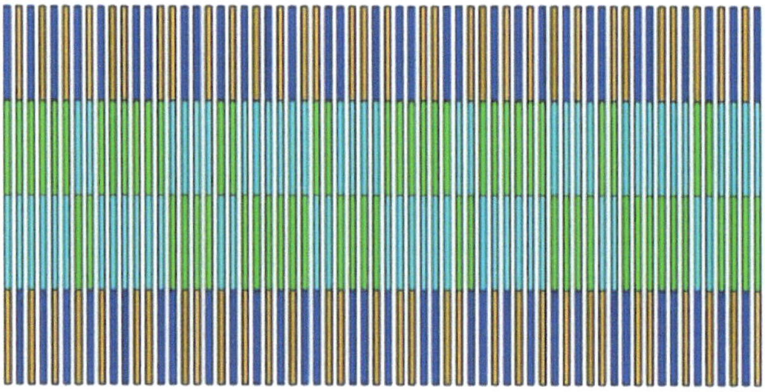

Figure 3.9: *This is an example of an encoded molecular structure that ultimately after a series of synthesis steps can form the DNA double helix, the base four genetic code, and the complementary planar steroid molecules, which will enable the operations of transcription, translation and replication. It represents the encoding of base four digital content into a molecular structure comprised as dimers of tetra-ring planar steroid molecules self-assembled into a long chain.*

Chapter 4

Transmitting DNA Structure and Function

To transmit the information enabled by the structural representation of the DNA molecule, it must be packaged in a stable configuration. Moreover, it must be accessible such that it can be decoded and used at the receiving end. In this chapter, the procedure to package DNA and its machinery necessary for its function is determined. The result is the DNA assembled as a double helix molecule, along with triggering agents, that is steroid molecules, which permit it to be accessed, decoded, and translated. In this section, to present the transmission of the encoded DNA complex, the following outline is applied:

- Description of the enzymatic complex that ultimately forms into an interleaved structure of complementary base paired DNA nucleotides matched to a planar steroid molecule of one of two types, either A-T complementary or C-G complementary.
- Synthesis of the products as arranged in the co-enzymatic complex format: these steps are largely described in Appendix A.
- Release of the final products in their final form of the DNA double helix, comprised of four complementary base pairs, and the planar steroid molecules that are necessary for decoding the DNA double helix, and for translating the encoded structures into a string of amino acids that form proteins.

4.1 Stacked Planar Steroid Assembly

The molecular structures of the planar steroids comprise a stacked array in which there is paired interaction of one molecule upon the other, such that one influences the synthesis of the other, and thus comprise a co-formation process. An example is depicted in Figure A.21(a), the stack is presented of ten planar steroid molecules for the creation of the five nucleotide pair of (T-A), (A-T), (G-C), (T-A), (C-G), starting from the bottom. The other five molecules will be associated with planar steroid molecules, which can be used for replication and transcription. Both will be developed in the same complex, and both are interactive and influence each other. If the phosphodiester chain covalently links to molecules 1, 3, 5, 7 and 9, the nucleotides will form from molecules 1, 3, 5, 7 and 9, and the planar steroids will form from molecules 2, 4, 6, and 8, with the numbering starting from the bottom. This example is studied further in Appendix A.3.

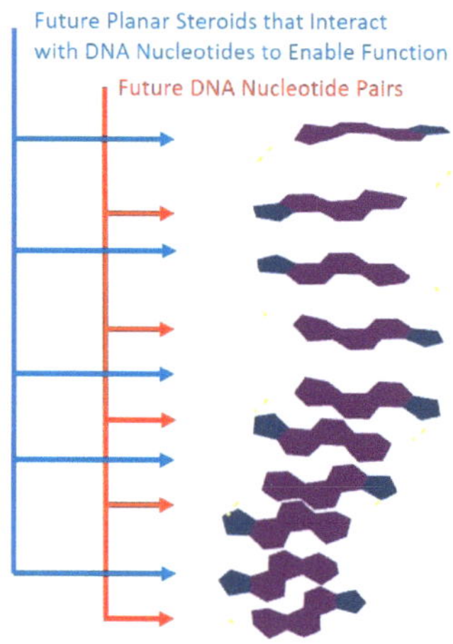

Figure 4.1: A stack of five dimers of which with a proper binding of the phosphodiester chain and interdimeric coupling during the synthesis process will result in five nucleotide pairs and five complementary planar steroid products. From this information, it is possible even at this stage using the representation of the ring orientations to determine the identity of the resultant nucleotides and planar steroids.

4.2 Synthesis of Unified Complex

Following the suggested reaction mechanism outlined in Appendix A, an initial step is represented in Figure 4.2 of linking the phosphodiester backbone, to the sequences, which is shown for the repeating AGCT unit, here shown as if T preceded A, or equivalently A followed T. The phosphodiester linkage is made through covalent bonding to the sugar on molecules 1, 3, 5, 7, ... and through hydrogen bonding to molecules 2, 4, 6, 8, ... of which the suggested reaction mechanisms indicate a perfect vertical spacing arrangement as with the DNA molecule.

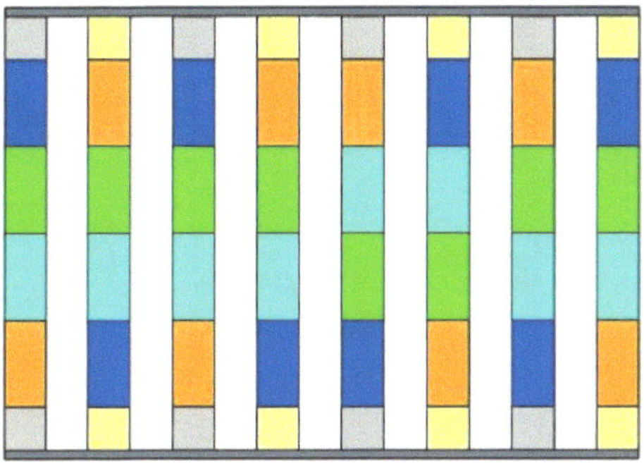

Figure 4.2: Binding a phosphodiester chain with covalent connection to both sides of the future DNA nucleotides while securing the accompanying steroid molecules through hydrogen bonding to the phosphate group achieves a unified complex that enables co-enzymatic activity in forming the DNA nucleotides and planar steroid product. Thus, the unified complex has sufficient structure to of achieve both structure and function for information storage and subsequent access.

The next step in the sequence involves the application of the interdimeric coupling to introduce differentiation and separation into the covalently linked planar steroid molecules, thereby forming the DNA nucleotides. This is illustrated in Figure 4.3, in which the orientation of the molecules is noted, and the nucleotides are noted through the color key described in Appendix A.7. The planar steroid molecules are also indicated by the color coding. The steps necessary to produce this complex are described in the Appendix. Here I describe the coding principles and transmission schemes, and thus the interesting details of the chemical reactions are included in the appendix.

Figure 4.3: The synthesis steps result in the DNA nucleotides separated into complementary base pairings, coupled through internal hydrogen bonding. Intermolecular hydrogen bonding is achieved through the AT complementary planar steroid and thymine.

4.3 Simultaneous Release to Form the Double Helix and Complementary Planar Steroids

Even though the DNA nucleotides have been formed, it is not in a format that is stable for transmission, and thus a method is needed to stabilize DNA. And by stabilizing the unit, which is accomplished through the formation of the double helix, both molecular structure via DNA and the molecular capability for useful activity are formed through the release of planar steroids, thereby instantaneously creating life function operations.

In the process of stacking dimers, and then subsequently bonding the arrangement according to a particular pattern, it is noted that some of the synthesized nucleotides are out of phase with respect to the nominal orientation, as shown in the illustrations of Chapter 2, which indicate that for standard DNA the purines and pyrimidines face in the same direction respectively. In the unified complex, they are of a mixed orientation, some dinucleotides are forward oriented, and some are backward oriented. Thus, there needs to be an orientation step of rotating these nucleotides, which are out of phase, to an in-phase configuration such that a stable structure of DNA can be formed. This structure is the double helix, and in the process, it will release the interleaved complementary planar steroid molecules during the rotation of the dinucleotides, such that the double helix can be formed.

It is seen in the illustrations of Figure 2.1(b) for the double helix that the second ring of the purine nucleobase is in a forward position, while the second ring of the pyrimidine molecule is in a back position, and this configuration needs to be achieved to stabilize the molecule. For a random sequence of nucleotide orientations within the unified complex, on average, half will be oriented in nominal position and half will be oriented in the opposite direction. The orientation in the nominal position consistent with the double helix, and the release of their complementary planar steroid molecules, is indicated in Figure 4.4, shown for the nucleotide sequence read as AGCT. This structure is stable as a double helix.

(a) DNA Double Helix (b) Released Planar Steroids

Figure 4.4: (a) Releasing through the formation of the double helix, which involves a re-orientation of the purine and pyrimidine molecules such that their orientation as seen along with one strand is uniformly opposite to each other. That is the pyrimidine are facing forward relative to the second ring, whereas the opposite is the case for the purine molecules. (b) The release of the planar steroid molecules through rotation is necessary to achieve the double helix. It indicates that there are an equal number of planar steroid molecules to that of the co-synthesized DNA nucleotides. There are two types of complementary planar steroid hormones produced, A-T complementary which has a hydroxyl group on the C-ring, and C-G complementary. (A-T and T-A complementary are equivalent; C-G and G-C complementary are equivalent.)

In the ejection of the planar steroid molecules, two types are formed, one type is A-T complementary and the other type is C-G complementary. These are represented in Figure 4.5 of the planar steroid molecules which have two types, one with an intramolecular hydroxyl group on the C-ring, and the other without one. These molecules are the precursors to steroid hormones, but for now, are used for other functional purposes.

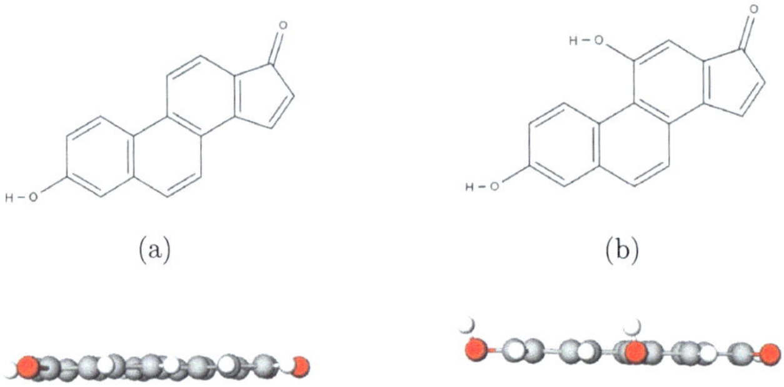

(a) (b)

(c) (d)

Figure 4.5: In addition to the DNA nucleotides, planar steroid molecules are a product of the synthesis steps associated with the unified complex. The planar steroid molecule of Figure 3.1(a) is the starting molecule, and the planar steroid product molecules have a major role in enabling life function along with DNA, in addition to providing catalytic and differentiating capability to produce DNA in the first place. There are two types of planar steroid molecules involved, and later released during the transition to the double helix are: (a) without a hydroxyl group in the C-ring, which is the same as the starting material, and the other (b) with a mid-molecule hydroxyl group. In (c) and (d), it is noted that both molecules are planar constructs and thus are consistent with the planar structure of the DNA nucleotides.

During the flipping process, some regions that are highly regular with alternating regular regions of TA and CG will not require flipping since they are already in position. Thus, these areas will stabilize the strands and there will be an opening. This provided a zone of stability, but the inability to move, and thus created small zones of stability about it, which resulted in the local separation of the strands. Immediately the strands were thus accessible and capable of an assembly process of amino acids as will be discussed in the subsequent chapters.

4.4 Strand Separation to Access DNA

While the DNA nucleotides are now in a stable double helix configuration after ejection of the planar steroid molecules by rotation of the out-of-phase nucleotides, in order for DNA to have purpose, it needs to be accessible for reading and action. It is good that the double helix is so stable, but it also limits its availability for opening and reading, and without that capability it is just an interesting polymeric molecule but would serve little purpose. Fortunately, it was co-synthesized with its key, the planar steroid molecule, that can open it. Therefore, in a transmission format, it is necessary to transmit not only the DNA molecule but also its decoding key to open it. Therefore, it is favorable for DNA, and its paired molecules, to be coupled in a cell format to contain the molecules. It is possible to deploy multiple planar steroid molecules to induce a stronger association and effective binding of the DNA strands, and thereby improve the duration of the subsequent opening of the strands of the nucleic acid. Thus, with intermolecular binding the strongest, it would favor chains of TA to form stabilization sites, or longer chains of the weaker CG interactions, which do not have an intermolecular bond with the planar steroid molecule. Strand separation is described in the Appendix through mathematical modeling.

Thus, having produced the paired and encoded structures of both DNA and planar steroid molecules, the bases will only be accessible for replication and transcription or translation if the strands are possible to separate. Fortunately, the planar steroid molecule is available, and can bind onto the phosphodiester chain to enable strand separation. This occurs through stabilization of the strands, which are vibrating and oscillating molecular constructs. Thus, there is kinetic energy transferred to potential energy through the stabilization process, whose internal energy must be dissipated through molecular disorganization, and bond breaking in this case. As the DNA molecule is stabilized by binding with the planar steroid molecule, there must be a balance in energy in terms of a release through bond breaking to be nearly equivalent to that of the stabilization provided by the binding event. Essentially, the DNA strands act as two strings in constant motion, which is fluidic, and with the cross-bonding induced by the planar steroid molecule, that is the triggering element, it acts as a pinch-point, and the strands, which were in relatively independent motion become constrained at one point, and around that point, a small bubble must form through hydrogen bond breaking upstream and downstream of the pinch-point. In Figure 4.6 the result of an open single-strand DNA is shown.

After encoding DNA as comprised of a set of planar steroid molecules with a specified orientation and binding pattern, their transmission as a sequence of DNA nucleotides arranged as a double helix, which was accomplished in conjunction with planar steroid molecules as part of a unified complex that is co-catalytic, with one molecule influencing the other to arrive at a final product set. Both molecules are useful as the DNA molecule provides the library, and the planar steroid molecule provides access to DNA, by separating the strands through a binding operation. The

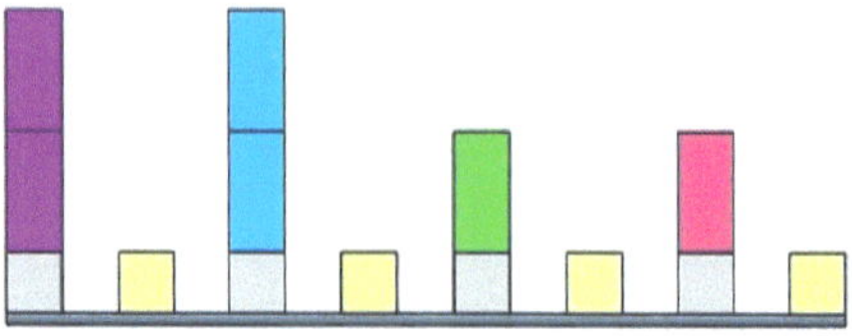

Figure 4.6: Opening of the double helix through the binding of planar steroid molecules at the initiation or promoter sites, thereby separates the strands and completes the transmission of the information process, as the message is received, therefore making available the DNA nucleotides for error analysis and decoding. The single strand of AGCT is represented here, connected through a single phosphodiester backbone and sugar.

dramatic nature of the rotation of the unified complex into the double helix thereby creates an instantaneous moment of the convergence of structure and function, and thereby a life function of a genetic purpose.

Chapter 5

Decoding DNA

Having formed DNA and its complementary planar steroid molecules and having the ability to access its code through the binding of the planar steroid molecules to the DNA strands, it is now necessary to decode it. In this chapter, the steps necessary to decode DNA are described. To differentiate this step from translation, I will take the decoding step up until the point where its interaction with amino acids occurs. With this chemical procedure, the prebiotic production of encoded proteins originating from nucleic acids can be precisely defined. The planar steroid molecule is a key component, as it is the starting material for the formation of the DNA molecule in a co-enzymatic framework, thereby encoding both itself and the DNA structure, and also effectively resulting in encoded structure for assembly of available amino acids.

5.1 Nucleotide Error Checking

In the context of initiating the decoding process after receiving the message, the planar steroid molecules, which are capable of self-assembled stacking, thereby align to each dinucleotide, and thus in effect testing the integrity of each nucleotide. This would involve the capability to mirror the nucleotide, including its presence. This is indicated in Figure 5.1, in which each steroid molecule is coupled with a corresponding nucleotide, depending upon whether hydrogen bonding to thymine is available. Thus, the planar steroid with a hydroxy group will pair with A-T and those without one will pair with C-G. In Chapter 8, error correction and damage control is discussed in which it is pointed out that the original orientation and code can be determined precisely by knowing the starting point orientation. Note that by error checking, it has the benefit of establishing a structure with a structurally uniform frame. As indicated in Figure 5.1, the strand opening provides access to the steroid molecules, and leads to an association with its planar steroid molecule. This process forms a sequence of alternating steroid molecules in the same height, with shorter nucleotide bases in between each uniform structured steroid molecule.

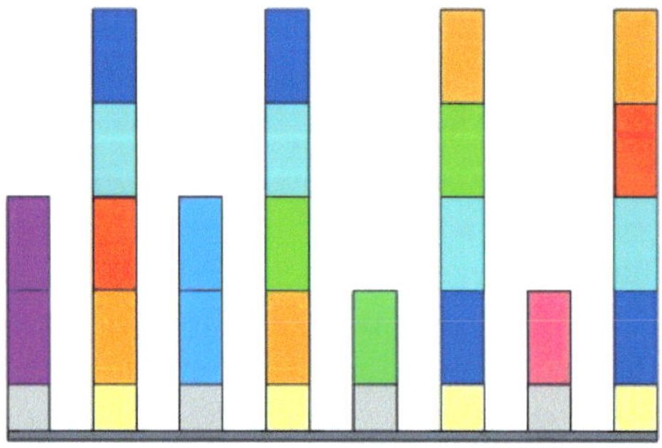

Figure 5.1: As an initial step after the message is received as transmitted, in starting the decoding process, error checking is performed on the nucleotides by aligned coupling to the complementary planar steroid molecules. The appendix includes additional discussion on error correction procedure, which is possible due to the structured encoding process.

As a basic result shown in Figure 5.12, the opening of the strand and association with planar steroid molecules through hydrogen bonding, enables a repeating self-contained interaction vessel as indicated by the dotted line. Within each set of three, there will be a zone which is comprised of two walls: one wall is the planar steroid molecule affiliated with nucleotide one, and the second wall is affiliated with nucleotide two. Moreover, each wall is supported by the nucleotide on the opposite side of the dashed enclosure, and thus the third nucleotide is needed to achieve structural symmetry of the zone. The third planar steroid molecule is complementary to the third nucleotide. Furthermore, the third wall, while it does not significantly contribute to the direct formation of the interaction vessel, it will serve as an alignment and docking post.

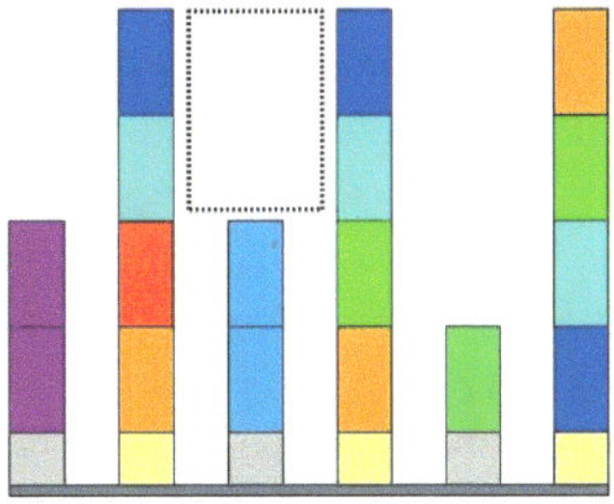

Figure 5.2: As a basic result, the opening of the strands and association of each nucleotide with a complementary planar steroid molecule, enables the construction of a repeating self-contained interaction vessel, as indicated by the dotted line. Within each set of three nucleotides, there is a unique zone which is comprised of three walls: the first wall is the planar steroid molecule affiliated with the first nucleotide; the second wall is affiliated with the second nucleotide, which together with the first wall define the "interaction vessel", shown in the schematic as a dashed line. Moreover, each wall is supported by the nucleotide on the exterior of the dashed enclosure, and thus the third nucleotide is needed to achieve structural symmetry of the interaction vessel, with the third nucleotide supporting the second wall. The third wall corresponding to a complementary planar steroid with the third wall serves as an alignment and docking site in the construction of a ceiling, which is also composed of a planar steroid molecule.

5.2 Decoding Table for the Interaction Vessel

Referencing the procedures described for encoding and transmitting DNA as a double helix formed as an intermolecular enzymatic complex comprised of planar steroid molecules, Figure 5.1, is a depiction of the pairing of the Adenine nucleotide with its complementary planar steroid molecule, the Guanine nucleotide with its complementary planar steroid molecule, the Cytosine nucleotide with its complementary planar steroid, and the Thymine nucleotide with its complementary planar steroid. Thus, the chemical decoding key is enabled, which is needed to determine the molecular interaction with the nucleotide and its matched steroid molecule. This will be the view when the strands of the double helix are separated, and the steroid molecules are matched up to the corresponding base, such that it is structurally and chemically compatible.

By combining the reference key of Figure 5.1 with all possible configurations of the triplet of nucleotide-steroid pairings according to the results of Figure 5.12, comprising the interaction vessel, it is seen that there are 64 possible configurations which can be arranged in table format of individual tables for Adenine in Figure 5.3, for Guanine in Figure 5.4, for Cytosine in Figure 5.6, and Thymine in Figure 5.5. Thus, a triplet nucleotide is selected, and the structural and chemical nature of its associated interaction vessel is determined. Because each interaction vessel is different, depending upon the sensitivity of the chemicals, it is possible to arrange for 64 different molecular interactions. In Chapter 6 on translation, I will show how each interaction vessel has a structural configuration that corresponds to the 20 standard amino acids. For the remainder of this chapter though, I will show how the protein is formed.

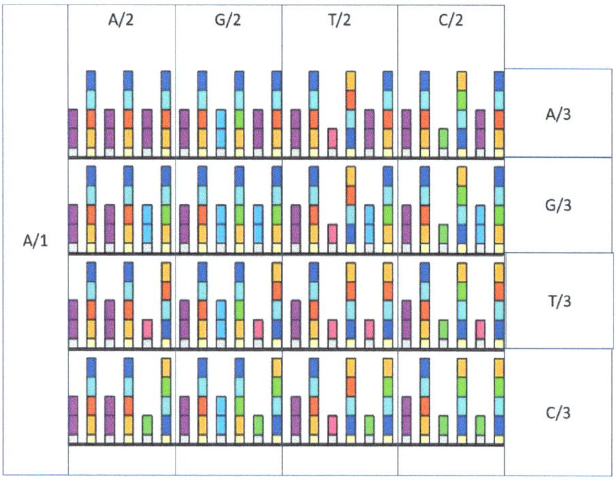

Figure 5.3: Starting with adenine, there are sixteen unique interaction vessels. Thus, the first wall is the same for each unit, but differs in the construction of the second and third sites.

5.3 Association of the Planar Steroid Molecule to Amino Acids

To enable the association of the amino acids to the interaction vessels defined by the nucleotides in conjunction with the complementary steroid molecules, the planar steroid molecules serve this purpose by forming the ceiling of the interaction vessel, and thus naturally create one open slot every three nucleotides since there is a one-to-one binding of the functional groups from ceiling-to-wall. Thereby, an

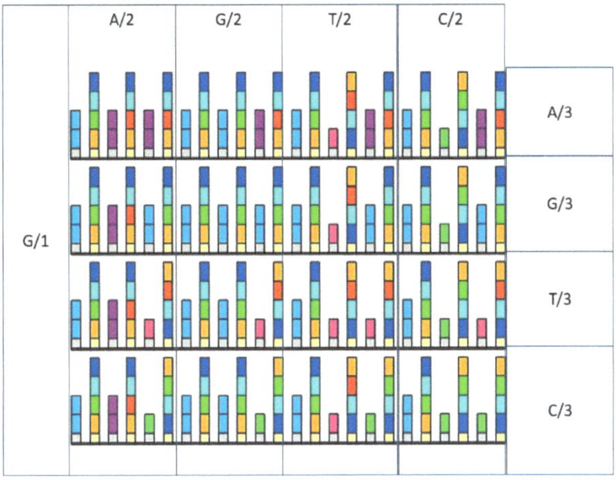

Figure 5.4: Similar to adenine, a unique set of sixteen interaction vessels is achieved that start with guanine. Like adenine, guanine is purine molecule.

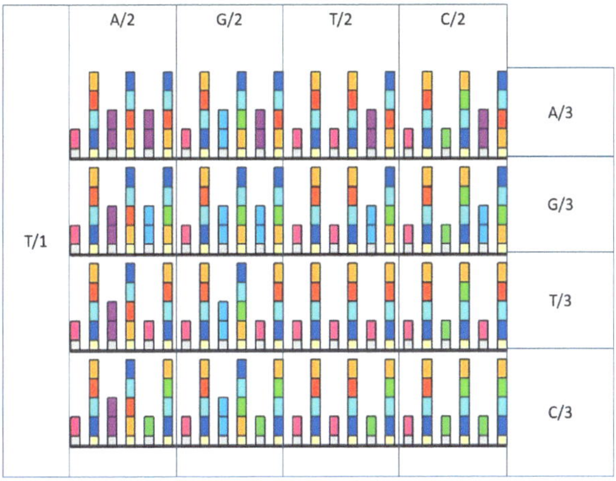

Figure 5.5: Another sixteen unique interaction vessels are unique that start with thymine, a pyrimidine molecule.

amino acid can bind to the opening of the interaction vessel and interact with the nucleobase and sides of the walls. The initial step involves a stabilization of the housing for the interaction vessel as a planar steroid molecule is laid across the top of the walls. This retains the planar steroid molecules as wall structures. Fortunately,

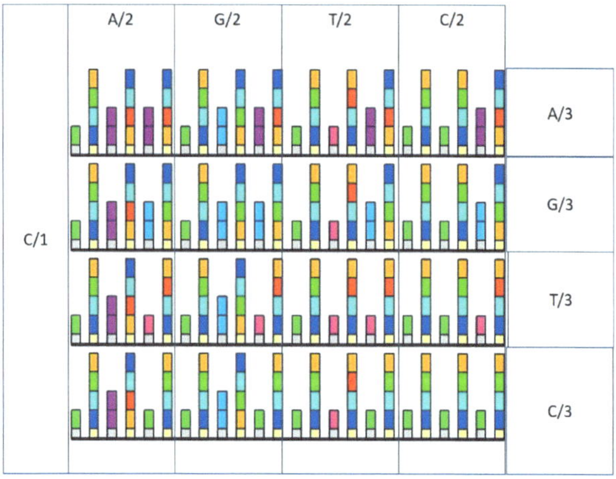

Figure 5.6: Completing the set of interaction vessels is cytosine, which also has sixteen unique interaction vessels. Thus, combined with adenine, guanine, and thymine, there are a total of sixty-four unique interaction vessels. The self-contained structure is situated above the second nucleotide.

the planar steroid molecule itself is a useful structure because it is indeed planar, and further will be available since it travels with the DNA molecule via the unified complex. Moreover, it contains two functional ends, which will permit multiple levels of interaction in structural construction and material stabilization through hydrogen bonding. Further the A-T complementary steroid molecule contains a third functional group in the mid-point of the molecule will permit integration with three consecutive walls.

Importantly, the planar steroid molecule can also interact with either end of the amino acid, as indicated in Figure 5.7. This planar steroid can then be used to position an amino acid within the vessel and connect to the next amino acid. Its size is perfectly suited for the interaction and has the advantage of blocking off the other two openings that are adjacent to the middle nucleotide opening. By associating it within a single direction, it further filters the amino acids, to just the L-amino acids, in order that the residues project down the interaction vessel.

This association of the A-T complementary planar steroid to form the ceiling of the interaction vessel is indicated in Figure 5.8. It is thus able to bind to the tops of the three walls through hydrogen bonding. The dimension is perfect to cover two openings and thus defines the structural reason as to why the codon consists of three nucleotides.

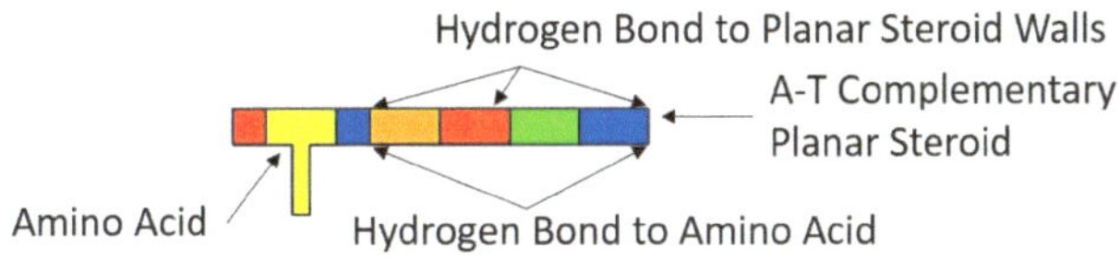

Figure 5.7: In forming the ceiling of the interaction vessel, the amino acid which stabilizes the housing in association with an A-T complementary planar steroid molecule. The planar steroid forms hydrogen bonds with the top of each planar steroid comprising the walls. There is overlap with the neighboring interaction vessel region. This structure repeats along the ceiling of the interaction vessel structures with different amino acids selected to optimize the stability of the structure based upon the structural and chemical aspects of the interaction vessel, which are discussed in the subsequent chapter.

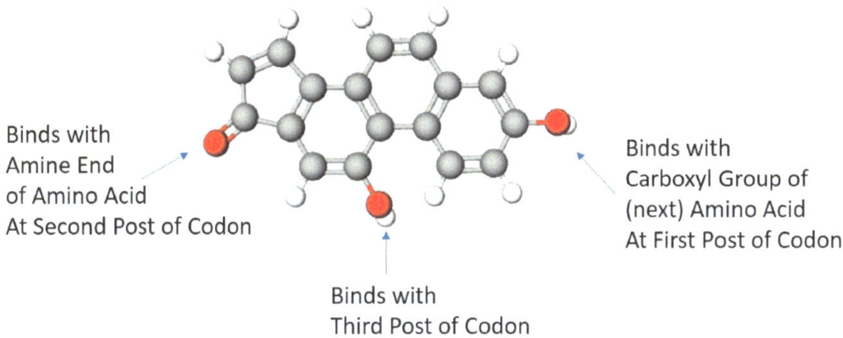

Figure 5.8: The chemical structure of the A-T complementary planar steroid, which also is used to form the ceiling of the interaction vessel and secure the amino acid. It exhibits hydrogen bonding with all three functional elements of the planar steroids comprising the walls of the codon.

5.4 Interaction Vessel for Amino Acid Assembly

Three consecutive nucleotides, together with the interleaved planar steroid molecules as walls, form an interaction vessel. With respect to the middle group, as indicated in Figure 5.9, the amino acid will interact directly with its two walls and its nucleobase. These two walls are in turn supported by the neighboring nucleotides on either side of the left and right wall. There is a third planar steroid molecule that does not form an interior wall of the interaction vessel, and is there because the third nucleotide is supporting the right wall, in a manner corresponding to that of the first nucleotide; the third nucleotide wall will also serve as an alignment and docking point for the ceiling of the interaction vessel, which is comprised of an A-T complementary planar steroid molecule. In addition, this interaction vessel also has aromatic walls, and has the possibility for a functional group as the hydroxyl group of the A-T complementary planar steroid molecule.

Note that a form of bitwise error correction could also take place, if the planar steroid molecule cannot orient properly it will not allow the interaction to proceed, thereby enabling time for a correction to be made to the error in question, such as the replacement of the proper nucleobase, or, if possible, the nucleotide base could be repaired or neutralized so that further interaction with planar steroid molecules could take place. The size of the interaction vessel and its opening would also provide a strong filtering mechanism to limit the type of materials entering into the unit, or if one is in place, its disassociation in favor of the planar steroid - amino acid complex that could stabilize the chain of interaction vessels.

To use the planar steroids in the transcription process, in addition to opening the strand through upstream binding, the planar steroids are associated with its corresponding nucleotide, and through the self-assembled stacking process, are organized as in Figure 5.9. This shows two codons, that is three nucleotides each, developed in an assembly interaction vessel. The planar steroid comprises the ceiling of the transcription vessel, and the amino acid interacts with the opening at the second slot of the three slots comprising the codon. The selectivity is based on the structural and chemical compatibility of the amino acid with the interaction vessel. This diagram is the basis for the codon to amino acid map, that is the genetic code.

The top view with a molecular model further documents this association of amino acid to planar steroid comprising the ceiling of the interaction vessel in Figure 5.10. The hydrogen bonding of one end is attracted to the amino group, and the hydrogen bonding is attracted to the carboxy end of the next amino acid. Thus, a codon is formed, and efficiency is established by preventing amino acids or other materials from occupying the adjacent slots.

5.5 Structural and Chemical Nature of the Interaction Vessel

The structural and chemical characteristics of the interaction vessel defined by the three nucleotides is further studied. In Figure 5.9, the structure of the interaction vessel over the second nucleotide has a major and highly sensitive influence over the type of amino acid that is incorporated within the interaction vessel, which is

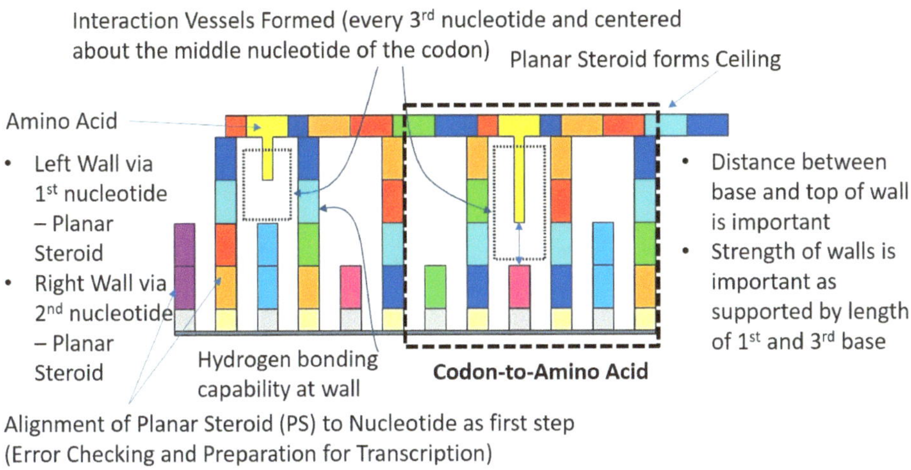

Figure 5.9: The transcription process starts with alignment of the complementary planar steroids with the nucleotides, which in effect is an error checking step. The next step is the formation of the ceiling via an A-T complementary planar steroid, binding to the top of the walls. An opening is established over every third nucleotide, allowing the entry of an amino acid to associate with the internal region of the interaction vessel. This relation results in the codon, the trinucleotide mapping to an amino acid. It answers the question on why the genetic code has three nucleotides coded per amino acid.

to be expected. The most significant factor is the distance between the top of the steroid molecule, which is the entry point for the residue of the amino acid and the top of the base. For the purine bases the distance will be greater than the pyrimidine bases. Other factors include the position of the oxygen group on the wall; the support provided by the adjacent nucleotides for the first and the third also play a significant role. The pyrimidine groups will provide less support than the purine molecules because of its height. On the other hand, entry into the vessel will be more hampered by the purine groups because the wall will not be able to move as much, therefore providing greater stability of the wall.

The chemical aspects are considered in the transcription framework diagram of Figure 5.9. The chemical interactions, like structural, have a significant role on the type of the amino acid that can enter and stabilize within the interaction vessel. Of primary importance is the chemical interactions with the nucleobase at the floor of the interaction vessel, if there is size compatibility, as well as the interaction with the

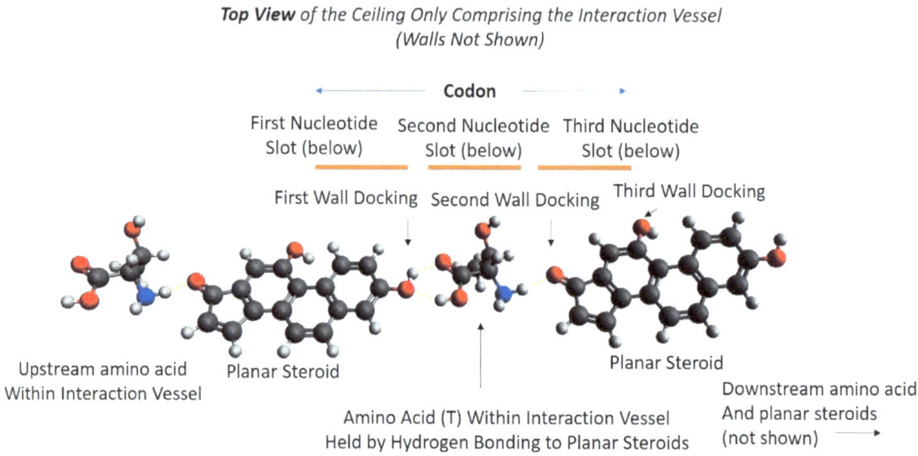

Figure 5.10: Top view of the ceiling of the interaction vessel is indicated. The walls are not shown, and are below the ceiling, secured to it by hydrogen bonding. The ceiling is defined by the A-T complementary planar steroid secured to the walls by hydrogen bonding. Every third nucleotide is open, and is thus occupied by an amino acid, thereby providing the reasoning on the relation of three nucleotides defining one codon per amino acid, pursuant to the genetic code.

functional groups associated with the walls of the interaction vessels. The chemical interactions with the first and third nucleotides are far less important but can play a role as it may pull the nucleotide away from the wall, and thus impact the structural support. Thus, the interaction vessel provides a differentiated interaction zone for chemical association. The chemical and structural characteristics of the interaction vessel are examined in the following chapter.

5.6 Association of Interaction Vessel and Amino Acids

This interaction of the vessel and amino acid is shown in Figure 5.11. Here the candidate amino acid is brought into the opening of the interaction vessel. If stability can be achieved through interaction of the candidate amino acid with the walls comprising the interaction vessel and the base, and if binding is significant, then the unit comprised of amino acid and planar steroid remains until removed due to elongation. If it is not compatible, then the amino acid will move along, and another candidate amino acid will try to lock into the interaction vessel.

The association of the amino acid and planar steroid wall can form chains since the two ends of the planar steroid molecule are functionalized, and both ends of the amino acid are also functionalized in a complementary manner. The ketone group of the planar steroid will interact with the amine group, and the hydroxyl group of the planar steroid will interact with the carboxyl group of the amino acid. The planar steroid will cover the two nucleotides that are not engaged with the amino acid and thus will make best use of the amino acid material. Furthermore, the planar steroid on comprising the ceiling of the interaction vessel structure will stabilize and enable the interaction of the amino acid with the walls and the base nucleotide to determine if a match is met. This can be illustrated in the example shown in Figure 5.11. An example of how it is put together for three amino acids, which require three interaction vessels and nine nucleotides, since each interaction vessel is located on the middle nucleotide of the triplet, in which the first and third nucleotides impact the structural and chemical characteristics of the interaction vessel. The connector blocks the third and first openings while the interactions are taking place in the second interaction vessel. The size of the connector which is a planar steroid molecule fits the configuration. When the interaction is stabilized, it rotates out of the vessel for attachment with the next amino acid in forming the protein through an elongation process.

Figure 5.11: An example of how it is put together for three amino acids, which require three interaction vessels and nine nucleotides, since each interaction vessel is located on the middle nucleotide of the triplet nucleotide, in which the first and third nucleotides impact the structural and chemical characteristics of the interaction vessel. The connector blocks the third and first openings while the interactions are taking place in the second interaction vessel. The size of the ceiling connector which is a planar steroid molecule fits precisely to complete the configuration.

5.7 Elongation of the Decoded Amino Acid Chain

Now having formed a stabilized structure, it is necessary to form the protein. In Figure 5.12(a), a mechanism is shown for peptide elongation. The planar steroid carries the growing peptide to the peptide for elongation. It forms a dimer, as indicated in the molecular model of Figure 5.12(b), with the ceiling and thus the growing peptide and the peptide to be bonded are in alignment. Further, the planar steroid molecule participates in this activity by alignment and by carrying any additional charge associated with the peptide bond formation. A molecular model of this event is given in Figure 15.6. This elongation process will continue until the stop site is reached.

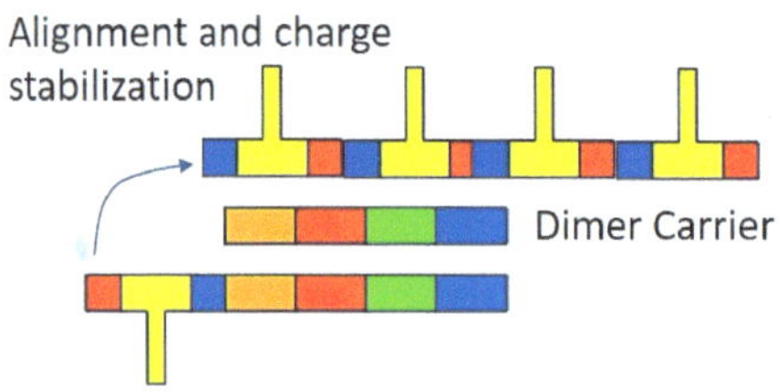

(a) Dimer Align

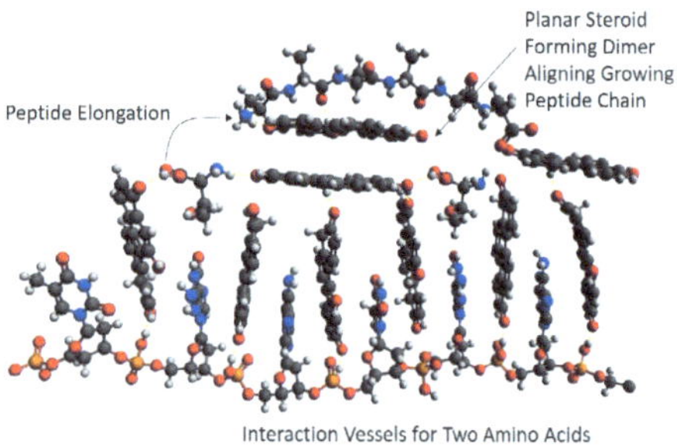

(b) Molecular Model

Figure 5.12: To produce a protein, after there is a period of stabilization dependent upon whether there is an adequate fit, the planar steroid enables the elongation process. (a) The planar steroid forms as a dimer, carrying the peptide chain and will drive the alignment and formation of the peptide bond. (b) A molecular model is used to demonstrate the dimer configuration that aligns and forms the peptide bond to the growing chain. Two interaction vessels are shown.

Another example is presented in Figure 5.13 for five amino acids in which planar steroid molecules form the walls and ceiling of the interaction vessel, all maintained through hydrogen bonding including the planar steroid comprising the celling to each wall. The stability and chemical interactions are defined through the triplet nucleotides, and thus provide a differentiated environment for repeatable assembly of amino acids, as discussed in the following chapter.

In Figure 5.14, the result is indicated. After the amino acids are coupled through the elongation process of rotation out of the complex aided by the planar steroid

Figure 5.13: As an example, five amino acids are coupled into the organized structure in which planar steroid molecules form the walls and ceiling of the interaction vessel, all constrained through hydrogen bonding. The stability and chemical interactions between the nucleobase and amino acid candidates are defined through the codon, and thus provide a differentiated interaction environment for repeated assembly of amino acids. The amino acids are matched to the interaction vessel based upon the structural and chemical correlation with the associated interaction vessel, of which there are 64 unique configurations.

molecule, a protein results whose characteristics are defined by the sequence of interaction vessel. While the detachment of amino acid occurs during elongation, the corresponding planar steroid can re-associate with the top of the walls comprising the interaction vessels, and the process continues while the strands are open until the transcription bubble closes.

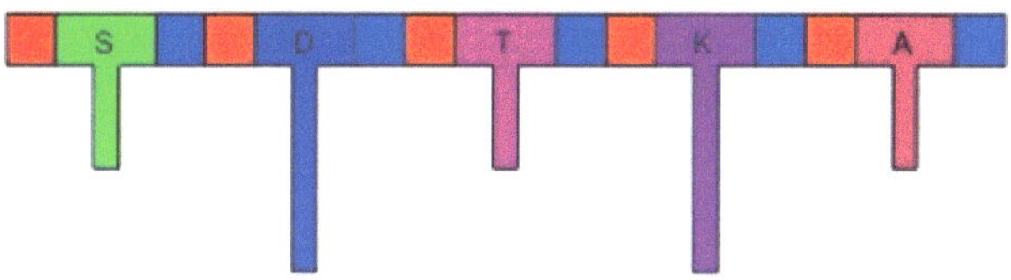

Figure 5.14: After the amino acids are coupled with the assistance of the planar steroid, a peptide results whose characteristics are defined by the sequence of interaction vessels from which stability within each interaction vessel was achieved. Thus, the process is completed and repeats as long as the strands remain open.

Chapter 6

Translating Decoded DNA

In the previous chapter, I showed how interaction vessels are formed and how amino acids assemble within the interaction vessels, separated by a planar steroid molecule which defines the ceiling and the walls of the interaction vessel. I showed how the spacing works, in which three nucleotides define an interaction vessel and there is a planar steroid molecule in between each assembled amino acid, thereby defining the map of triple nucleotide, that is a codon, to amino acid. In this chapter, I examine the chemical and structural sensitivity requirements of the interaction vessel to uniquely assign an amino acid to an interaction vessel. Here, I will discuss the feasibility of using the interaction vessel to differentiate amino acids, and mainly describe it from an inclusionary perspective, that is how the individual interaction vessels can support the intended amino acid pursuant to the amino acids known to form peptides of living cells. A translation chart is shown for the first time that maps each amino acid to a structural sequence that defines the basis for codons comprising three nucleotides.

6.1 Structural Sensitivity of the Interaction Vessels

To evaluate the sensitivity of the interaction vessels to differentiate between amino acids, it is useful to compare the structural and chemical characteristics of the interaction vessels formed by three walls of the planar steroid molecules and its associated nucleotide. For a set of twenty amino acids, in order to get to a unique set, it is necessary to find at least twenty unique differences in the structural and chemical composition from the set of sixty-four interaction vessels, which can be a combination of attributes. In evaluating the structural attributes, Table 6.1 indicates such significant differences among the different interaction assembly vessels.

The most striking comparison is the size of the residue group of the amino acid.

	Metric	Amino Acid																			
		A	C	D	E	F	G	H	I	L	K	M	N	P	Q	R	S	T	V	W	Y
Structural	Residue Oriented to Project Down the Vessel	X	X	X	X	X	X	X	X	X	X	X	X	X	X	X	X	X	X	X	X
	Shorter Residue	X	X				X		X	X		X		X			X	X	X		
	Longer Residue			X	X	X		X			X		X		X	X				X	X
	Linear Residue	X	X								X	X					X				
	Symmetric with Ring Structure					X								X						X	X
	Compatibile with Weak Walls Unable to Retain Residue						X														
	Larger Element than C, O		X									X									
Chemical	Oxygen and Amino End Groups	X	X	X	X	X	X	X	X	X	X	X	X	X	X	X	X	X	X	X	X
	No Functional Group	X					X		X	X									X		
	Functional Group Interaction		X	X	X	X		X			X		X		X	X	X	X		X	X
	Oxygen Influence Higher Position								X			X					X		X		
	Oxygen Influence Medium Position								X	X		X	X		X			X			
	Third Base Influence			X	X			X					X		X	X					
	Base Influence			X	X			X		X			X		X	X					X

Figure 6.1: The interaction vessels define a unique set of sixty-four zones with which to interact with the amino acid residuals. An initial analysis of the set of twenty amino acid to associate unique structural and chemical characteristics for each amino acid, which can then be related to the characteristics of the interaction vessel so as to assign an amino acid for each interaction vessel.

It can be seen in looking at a chart of the twenty amino acids that there is good correspondence between the assigned triplet and the associated interaction vessels. (Note, when performing this analysis, it is important to remember to swap out the complementary base pairs of a nominal UCGA table, since the approach here is to directly translate off of the nucleotide base, rather than go through the intermediary of mRNA). The shorter residue groups such as those of valine, alanine, threonine are associated with the bases that comprise the purine molecules, which will be closer to the edge since they are two rings, while the pyrimidine molecules contain only one ring.

Further structural issues include the strength of the sidewalls. At an extreme, the CCC pairing will be the weakest since there is little support, there are no functional groups on the sidewalls, since the walls are all CG-complementary without hydroxyl groups, to assist with stabilization, and there is a relatively long distance from the opening to the nucleotide base, cytosine. It is thus the case that no amino acid with a residue can enter the interaction vessel because it is too unstable and will not allow sufficient time to interact with the candidate amino acid. The only amino acid that can interact with CCC would be glycine, which has a hydrogen element as a residue group.

The chemical interactions of the vessel are also important and in Table 6.1, the chemical interactions are evaluated. The chemical differences include the wall functional group for the AT complementary planar steroid molecule, and its positioning.

Other chemical features include the nucleotide base itself, and whether there are two or three functional groups, and the nature of those functional groups, whether it be a ketone or amine group. The chemical charge of the bases on the first and third nucleotides is less important, but can play a role in certain circumstances, because it can strengthen the support of the first or second walls, or impact its potential chemical attractive forces.

It is important to comprehend the specificity enabled by the first and second walls corresponding to the first and second nucleotides, which comprise the interior walls of the interaction vessel, and thus, the translation code should emphasize the first two letters because of these requirements. The third amino acid is necessary to support the second wall, but is not as important as the first two nucleotides, as it is only providing symmetry and support, and serves as a docking point for the AT complementary steroid molecule comprising the ceiling. Therefore, the third nucleotide is necessary but not as much as a definitive measure of the performance of the interaction vessel since the third wall is on the exterior of the interaction vessel. This can be seen in the amino acid to nucleotide code, where the third nucleotide for some pairings is not influential at all, and for other interactions it plays a role. However, the third nucleotide is critical because it provides the wall from which the planar steroid molecule associating the amino acid to the interaction vessel performs alignment and orientation. Thus, a strong result indicating that this approach is correct is that the organization of the amino-acid to triplet nucleotide is explained in that the third nucleotide has less impact on the mapping of the amino acid to nucleotide, since it provides support and alignment, but not the actual wall for the association of the amino acid to the base.

Some unusual cases can be examined, such as the starting site, which has only one possibility. This is rationalized by considering the size of the SH group and its position within the candidate vessel, and the strength or flexibility of the side walls. This can be contrasted with the stop sites.

6.2 Translation Codes

In Table 6.1, amino acids with shorter residuals are analyzed for interaction with the purine nucleobases, which have two-rings and are thus larger and longer than pyrimidines to accommodate the shorter amino acids. The chemical characteristics of these amino acids, as well as structural characteristics, are evaluated to determine the characteristics of an interaction vessel which would enable a more stable configuration. The mapping is used when the interaction vessel and the adjacent nucleotide/planar steroid pair are given, and a prediction on the amino acid is needed. This analysis is provided in an inclusionary sense, which means that perhaps multiple amino acids would satisfy the properties of the interaction vessel to achieve stability. Additional work is needed, more modeling, to develop the table further in an exclusionary sense. But to a first order approximation, there is agreement, particularly on size.

AA	Interaction Vessel Requirements	Sequences
F	Symmetric, Functional groups, Strong walls to support aromatic ring	AAA,AAG
T	Functional Group positioned left low to attract OH	TGA,TGG,TGT,TGC
S	Functional Group positioned left higher to attract OH	AGA, AGG, AGT, AGC, TCA, TCG
M	Two functional groups to lodge large molecule with flexible walls	TAC
I	Charged walls at low level left, higher right	TAA,TAG,TAT
L	Strong walls, left charge positioned low	AAT,AAC,GAA,GAG,GAT,GAC
V	Weak left wall, no left functional group, two functional groups on base	CAA,CAG,CAT,CAC
A	Weak left wall, no left functional group, three functional groups on base	CGA, CGG,CGT,CGC
P	No charge, strong walls to support ring group, no functional groups	GGA,GGG,GGT,GGC
G	Weak walls prevent entry other than hydrogen,non-functional group	CCA,CCG,CCT,CCC

Table 6.1: For the purine nucleobases, which are longer molecules because of its two rings, the length of the interaction vessel will limit the kind of molecules that can stably interact within the vessel while retaining connection with the exterior of the vessel. Thus, the differentiation of the amino acids with physically shorter residues are compared.

As size is a good first indicator of the nature of the interaction between amino acid with the interaction vessel, Table 6.2 provides a the characteristics of the amino acids with the longer residuals, which would associate with the pyrimidine nucleobases, which have just a single ring and are thus shorter than purine, but can accommodate the larger sized residues of the amino acids. Glycine is an exception since it has a hydrogen as a residue yet associates with the pyrimidine base cytosine. This will be discussed further, but the basic reason is that the interaction vessel for CC as the first two nucleotides of the codon is unstable, and the longer residues cannot reach the nucleobase without distorting the vessel. Hence CC is skipped over via glycine.

In Figures 6.2, 6.3, 6.4, and 6.5, the decoding tables are shown that correlate

AA	Interaction Vessel Requirements	Sequences
R	Strong Walls, Uncharged walls, Three group interaction with C	CCA,CCG,CCT,CCC
R	Functional Wall on left for interaction	TCA,TCG
S	Functional Wall with O well positioned to attract OH	TCT,TCC
W	Some symmetry, stronger wall, some flexibility	ACT
C	Stable association environment with SH group, stronger walls, able to retract due to shortness	ACA,ACG
Y	Charged walls on both sides, and 2 functional groups	ATA,ATG
H	Strong walls, charge on right side wall	GTA,GTG
Q	Significant charge and stronger walls	GTT,GTC
N	Significant charge but weak walls, though not as weak as K	CGA,CGG
K	Significant charge but weak walls hence linear group	TTT,TTA
ZW	Larger molecule with stability	STOP
ZC	Stronger larger molecule with stability	STOP

Table 6.2: For the pyrimidine bases, which are shorter than purines as there is only one ring, the amino acid residues need to be relatively longer in order to interact and secure successfully while retaining connection with the exterior of the interaction vessel. The table examines the amino acids with physically longer residues in terms of other differentiating factors. ZW and ZC denote amino acids that act as stop sites. Note that there is a limit however to the length of the amino acid residue, because if the residue is too long, it cannot secure successfully to the nucleobase and to the entry point.

the structure and chemistry of the interaction vessel with the corresponding amino acid. The justification for these amino acids is taken from Table 6.1 and 6.2, and the structural and chemical checklist. It would be possible to correlate a second table of amino acids that may compete effectively for the interaction sites. This can then be inverted to determine the best amino acid given a set of conditions corresponding to the chemical and structural characteristics of the interaction vessels.

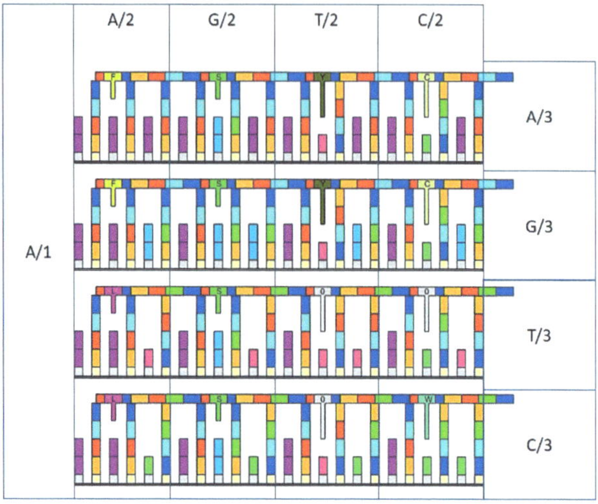

Figure 6.2: Mapping the construct to an amino acid for adenine as the leading nucleotide, the translation of the triplet structure to assemble an amino acid based on structural and chemical interactions over the second nucleotide. Note that the table can be organized on size relative to the second nucleotide, that is the columns of the table. This is consistent with the length of the interaction vessel being the first differentiation factor. An amino acid is assigned to each of the sixty-four interaction vessels, which is depicted by a letter stamped into the colorized amino acid. These selections are based on the tables of chemical and structure correlations, and on reflection of the amino acid to triplet nucleotide mapping.

6.3 Novel Translation Table

Thus, translation occurs through an interaction vessel assembly constructed from three complementary planar steroid molecules comprising walls and three nucleobases, in addition to two A-T complementary planar steroid molecules as the ceilings with half of each overlapping with the adjacent codon zones: the amino acid completes the construction process, entering into the open single slot centered above the second nucleobase of the codon. The structural and chemical sensitivity of the interaction vessel determines its association with a compatible amino acid. The orientation of the projection of the entry point of the interaction vessel provides a first level filter of the type of chemicals that can be associated within the interaction zone, which would be limited to L configuration as the D configuration will project in the

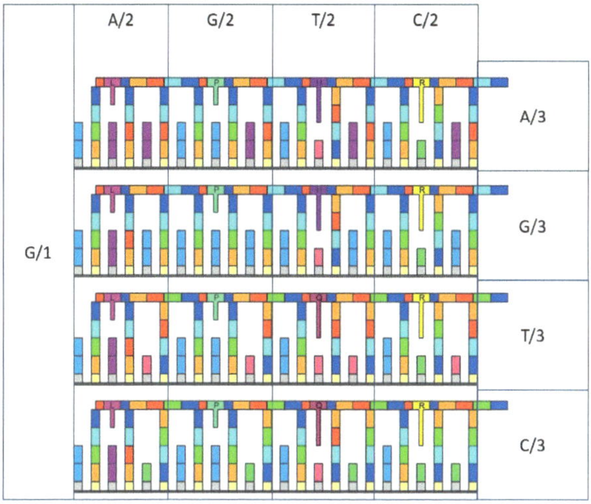

Figure 6.3: With guanine as the leading nucleotide and planar steroid structure, the decoding of the triplet nucleotide structure with an amino acid is indicated. The amino acids are colorized according to the color key and a letter is stamped into each amino acid. As a reminder, each rectangle relates to a different ring structure, whether on the nucleotide or the planar steroid molecule.

opposite direction. The length of the residue is the next factor as it immediately separates the amino acids into those that interact with purine and those that interact with pyrimidine due to the distance between the entry point and the number of rings comprising the nucleobase. More detailed chemical and structural characteristics are then used to determine the association of a set of twenty amino acids to the interaction vessels. Using the inversion tables, it is possible to predict which interaction vessel, and thus triplet nucleotide codon, corresponds to a candidate amino acid.

In Figure 6.6, an initial evaluation of the structural and chemical characteristics of the interaction vessel with the amino acid is indicated. The table indicates that there is a correlation in the physical size of the residue of the amino acid as considered in the first two columns associating with residues that are physically shorter than those of the third and fourth column. This is correlated to the size of the base where the first two columns correspond to a purine molecule which is two rings in comparison to the pyrimidine molecule which has only one ring. Thus, longer residues will be needed to interact and stabilize with the base in the third and fourth columns than those of the first two columns. Because of the longer distance, there

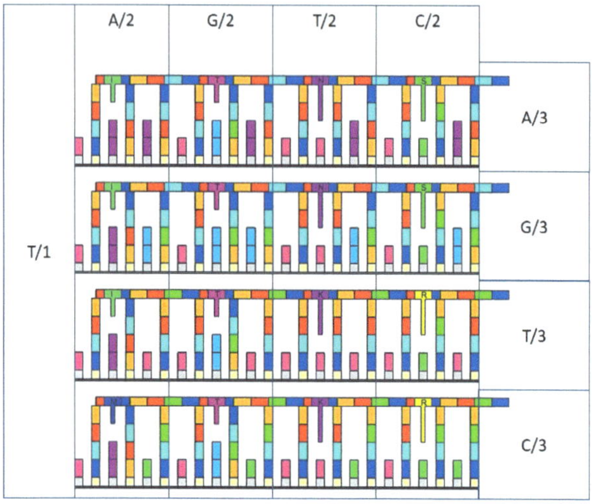

Figure 6.4: The decoding of the triplet with thymine as the first nucleotide, and as noted with the others, the second nucleotide is a major differentiating factor since it defines the length of the interaction vessel. In this section, the start sequence methionine is located at an individual site, which can be rationalized through examination of the chemical structures.

is an increased sensitivity to the strength of the walls, including the relative size of the base of the third nucleotide. A shorter base of the third nucleotide position will allow the wall to flex more and can lead to larger residues, which is the case in comparing Group A and B residues. Furthermore, glycine is the shortest amino acid, and the reason for that is that its interaction vessel is unstable, and nothing other than the hydrogen residue of glycine will work. In effect glycine skips over the unstable interaction vessel to maintain the integrity of the alternating and growing chain of amino acids and planar steroid molecules. The instability of the interaction vessel results because: (1) there are no functional groups on either wall and thus there is no opportunity to stabilize the interaction vessel through association with an amino acid; (2) there is a long distance from the entry point of the interaction vessel to the smaller pyrimidine nucleobase; (3) the external support provided by the first nucleobase is low since it is also the shorter pyrimidine and there is thus a lack of aromatic stabilization of the second ring comprising the first wall. Other modes of inclusion can be noted for this set of amino acids relative to the physical dimension and chemical characteristics of the interaction vessel. It is noted that this translation map is a first pass analysis and subsequent research will provide a more

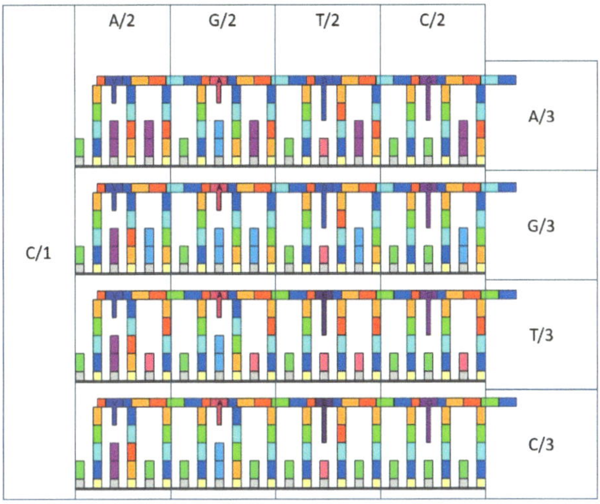

Figure 6.5: Completing a total of sixty-four interaction vessels for twenty amino acids, cytosine as the leading nucleotide is shown for the translation table. Of interest is glycine, which is the shortest amino acid residue, but it is present in the longest distance from entry to nucleobase interaction vessel. However, this interaction vessel is weakest since there are no functional groups on its sidewalls, the support on the side walls is weak from the short first nucleotide, and the distance is long. Thus, longer molecules are not able to stabilize successfully, and since glycine does not have these problems since its residue group is a hydrogen element, glycine is selected as the best at skipping over the unstable interaction vessel.

specific understanding, including the potential for amino acids outside the set of twenty to satisfy the chemical and structural requirements necessary for stabilizing within an interaction vessel.

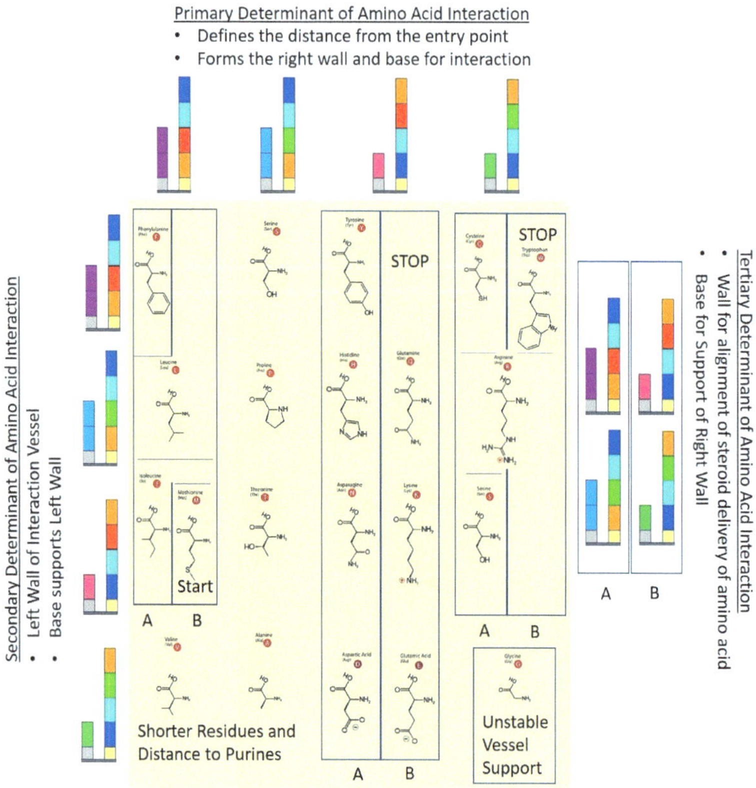

Figure 6.6: *The first derivation is shown of the genetic code mapping the construct of the interaction vessel to amino acid. Each segment is defined as a nucleobase and its complementary planar steroid comprising one of the three walls. The first nucleotide/steroid wall is indicated by the rows, the second nucleotide/steroid wall is indicated by the columns, and the third nucleotide/steroid wall is indicated by the Group A or B of a split column. Note that the size of the amino acid residue is generally correlated to the size of the distance between the tip of the planar steroid comprising the wall and the nucleobase positioned on the strand.*

Chapter 7

Example of DNA Encoded, Transmitted, Decoded, and Translated

Applying the results for the simulation and design of DNA for producing molecular structures, and in simulating what would be the origin of DNA and its transcriptional mechanisms, this section presents the chemical structures associated with encoding DNA, transmitting the information, and then decoding the information so as to obtain a desired result, which in this case is a protein structure. In addition, the inverse direction is discussed, in which the desired structure is specified, and then the encoded structure is synthesized, which would result in the desired structure when implemented.

Rather than specify the promoter as an initiation site to enable strand separation, a starting nucleotide triplet is specified to illustrate the encoding, transmission, decoding and translation. A random number generator is used to select the molecular orientations (that is, the arrangement of the four rings) for 25 pairs of dimers of the starting material, the planar steroid molecule; hence a total of 50 planar steroid molecules will serve as the initiating set of planar steroid molecules, which after synthesis will result in 25 nucleotides and 25 complementary planar steroid molecules. This structure will then be decoded into a series of interaction vessels comprised of a nucleotide triplet, codon, in association with the planar steroid molecules constructing its walls. Translation of this decoded set into a chain of amino acids takes place by association of an amino acid into the openings of the interaction vessel created by the A-T complementary planar steroid molecules comprising the ceiling of the interaction vessel. The identity of the amino acid is determined by the structural and chemical characteristics of the interaction vessel.

7.1 Encoded DNA

For this example, twenty-five dimers (a total of 50 molecules: see Figures 3.3(a), 3.3 (c), 3.3 (d)) of the planar steroid molecule of Figure 3.1(a) were selected independently by a random number generator with the sample drawn under uniform probability distribution from the set of orientations depicted in Figure 3.4. The structural sequence for the 50 molecules is listed in Appendix C, and is illustrated in the graphical form of Figure 7.1. The specification of this encoded structure as an interleaved set of nucleotides and planar steroids sets forth the entire synthesis process that will result in an amino acid chain and thus protein. (At the end of the chapter, I consider the reverse problem of specifying the nucleotide or amino acid sequence and determining the dimer orientation sequence which would result in the specified nucleotide or amino acid sequence.)

As there are 25 pairs of coupled molecules, there will be 24 inter-dimeric couplings. If the last molecule is considered as coupled with the first molecule, there will be twenty-five inter-dimeric couplings, and that is the route that will be taken to bring the example to completion. The orientation of the rings between the interdimeric coupling of the opposing molecules will define the nature of the resultant nucleotide, whether it will become A-T/T-A or G-C/C-G, as well as the complementary planar steroids, that is whether it becomes A-T or C-G complementary. Applying the encoding Table 3.2, the capacity for hydrogen bonding of the interdimeric molecules, which determines whether the resultant nucleotide is destined to be of adenine and thymine, is indicated as:

$$H_{bond} = [0111110010101010010001100] \tag{7.1}$$

Thus, this step of coupling the interdimeric molecules of planar steroid molecules effectively encodes a structure that will ultimately become a DNA nucleotide sequence as well as its complementary planar steroid molecules that will enable its information content as a life function. The message as a molecular structure is thereby fully encoded and ready to be transmitted. It is noted that at this stage the encoding of the molecule can be deciphered mathematically to determine the nucleotide sequence after the hydrogen bonding sequence is determined. From the orientation of the rings and the resultant hydrogen bond arrangement, the nucleotide sequence for this example is:

$$\mathtt{DNA}_{\mathtt{example}} = \mathtt{GAAATAGCACTGTCTCGTCGGTTGG} \tag{7.2}$$

Figure 7.1: The sequence of the orientation of the dimers comprised of planar steroid molecules is indicated for the twenty-five pairs. This figure is a representation of a physical structure, that is an arrangement of planar steroid molecules, and the coloring indicates the orientation of the ring. The complex is comprised of pairs of planar steroid molecules connected by hydrogen bonding on either end. Depending upon the orientation of the four rings comprising the planar steroid molecule, of which each ring is represented in the figure by a rectangle, the resultant nucleotide sequence is effectively encoded within this structure, as is the product set of complementary planar steroid molecules.

7.2 Transmitted DNA

To prepare the DNA and planar steroid co-enzymatic complex for transmission, it is arranged with an attachment of a phosphodiester backbone in which the DNA nucleotides are covalently attached. The interleaved planar steroid molecules are not covalently bound but are stabilized by hydrogen bonding to the phosphate group. This is represented in Figure 7.2.

This co-enzymatic complex defines a set of DNA nucleotides that are synthesized with the interleaved planar steroid molecules, in which one molecule from the dimer influences the opposing dimer molecule from the adjacent dimer. In this configuration, the result is that the DNA nucleotides form a connected base 4 construct, and

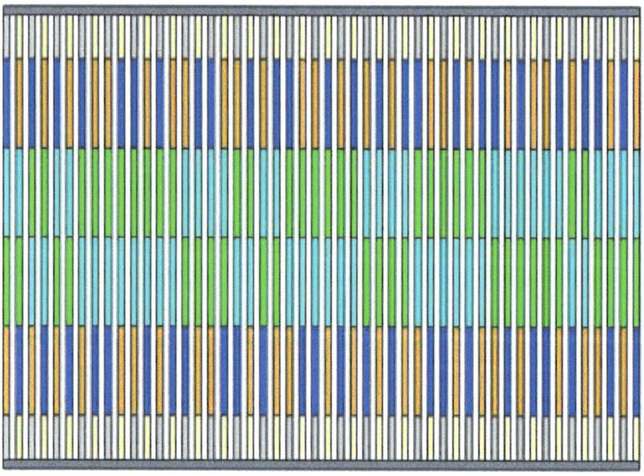

Figure 7.2: For the example, after association with the phosphodiester backbone the future nucleotides are attached by a covalent bond, but still comprise a planar steroid molecule, and the accompanying complementary planar steroid molecule are stabilized at its ends by hydrogen bonding to the phosphodiester chain, and internally by aromatic stabilization.

the planar steroid molecules take on two forms, either with or without a mid-molecule hydroxyl group, that is A-T complementary or C-G complementary. The steps of the synthesis are represented in the appendix, and using the procedures described, the formation of the unified complex prior to the formation of the double helix is given in Figure 7.3. This structure includes the nucleotide pairing and its associated steroid molecule. The color key is indicated in Appendix A.7. One important point is that some of the DNA nucleotides are not oriented in a position compatible with the double helix as we understand it, in which along the 5' strand the purines face forward and the pyrimidines face backward. In the unified complex, on average half of the pyrimidine and purine dinucleotides will be not be in the proper orientation, facing the wrong direction.

To form the double helix, it is necessary for the DNA dinucleotides that are not in proper orientation to be rotated into position, which will thereby release the planar steroid molecules from the complex. After nucleotide rotation, each purine is aligned in the same direction, and each pyrimidine is aligned in the same direction. Inter-

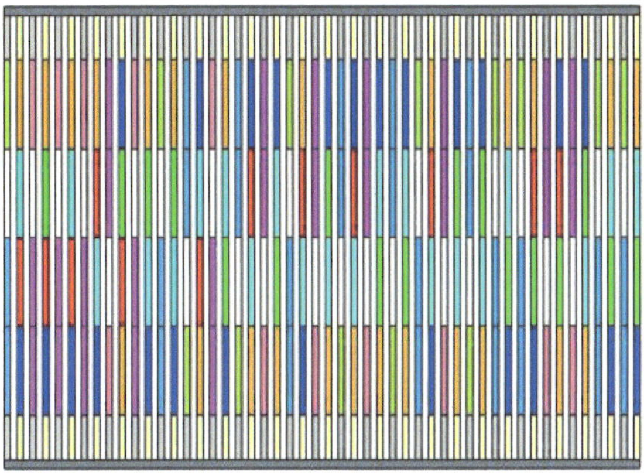

Figure 7.3: The unified complex is shown indicating that the nucleotides are synthesized, connected by to the phosphodiester chain by covalent bonding, and the complementary planar steroid molecules are in place with two types, either with an oxygen group on the C-ring for intermolecular hydrogen bonding to thymine, or without, equivalent to the planar steroid starting molecule. All structures have planar arrangements. Some of the purine and pyrimidine nucleotides are oriented the opposite way to which they face in the double helix, and thus the structure is relatively unstable in comparison to the destined configuration of released planar steroid molecules and the double helix of DNA.

estingly, the purines are aligned as BFBF and the pyrimidines are aligned as FBFB, which will balance the configuration. On average, this will require "flipping" half of the synthesized dinucleotides to obtain the stable structure of a double helix. This rotation frees the planar steroid molecules, which can then be used immediately to open the strands for further processing, including amino acid assembly and transcription. The nucleotides which would have to flip can be determined mathematically, and that sequence is given by for this example as

$$\mathtt{O}_{\mathtt{flip}} = [0101011100110011100001000] \tag{7.3}$$

in which $\mathtt{O}_{\mathtt{flip}}$ indicates whether the nucleotide flipped in order to arrive at the final configuration of the double helix. Thus, in this example, there are 11 nucleotides that rotated into position.

In Figure 7.4 the double helix is shown, which is obtained from the unified complex by rotating the out-of-phase dinucleotides and ejecting all planar steroid molecules. The strands are thus available for association with the same planar steroid molecules such that strand opening can initiate transcription. Thus, there is an instantaneous life function with the release of the planar steroid molecules, which interact directly with DNA to induce genetic function and distribution of the information content.

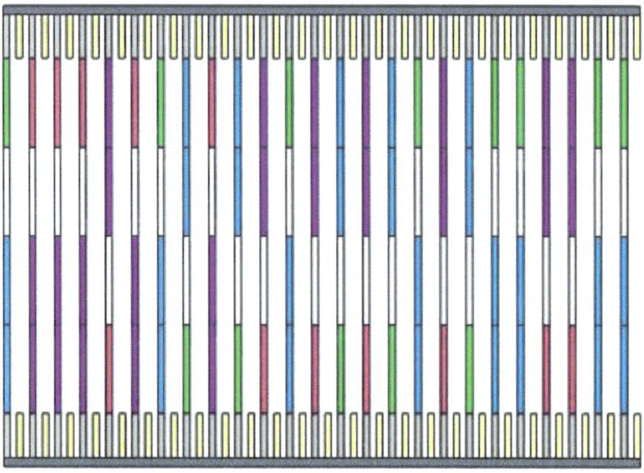

Figure 7.4: The formation of the double helix is obtained by rotating the nucleotides that are not in the nominal orientation corresponding to a double helix. This rotation results in the ejection of the interleaved planar steroid molecules. After ejection, the planar steroid molecules initiate transcription by binding to promoter sites on DNA.

7.3 Decoding DNA

After receiving the message as a separated strand of DNA, enabled by the planar steroid molecule inducing a transcription bubble by binding effectively at a promoter site, the next step is to perform error checking by a bitwise comparison with its associated steroid molecule in populating the strand at the location of the phosphor molecule on the phosphodiester backbone. The result of the nucleotide matched with a steroid molecule in its adjacent location is indicated in Figure 7.5, which is achieved by a self-assembled stacking process. In addition to error checking, this configuration prepares the overall structure for amino acid assembly by delivery of the amino acid by the planar steroid molecule, which also forms the ceiling of the overall structure comprising the interaction vessel.

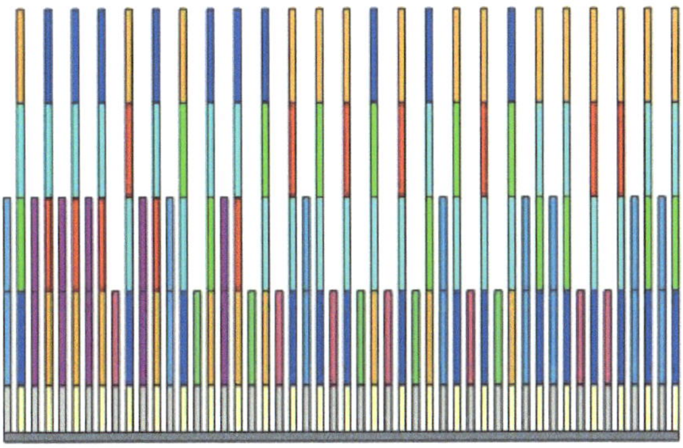

Figure 7.5: Separation of the strand through the initiation by planar steroid molecules, and then alignment of the complementary planar steroids with the exposed single strand nucleobase prepares the structure for transcription. Note that the planar steroid molecules when positioned in alignment with the nucleotide provide a uniform upper surface which will define a site where the ceiling to form an interaction vessel to secure amino acids for peptide elongation.

In Figure 7.6, the final configuration in which the matching of the candidate amino acids within the assembly vessel. The beam on top is comprised of planar steroid molecules which assist in stabilizing the candidate amino acids into the interaction vessel, and in blocking the adjacent potential slots. If in testing a candidate amino acid the interaction vessel is unstable or non-active, for example, the residue of the amino acid is too big or too small to fit within the interaction vessel or fails to bind because of a lack of functional modes, the process effectively moves onto the next amino acid candidate. If the fit between the residue is consistent with the chemical and structural characteristic of the interaction vessel works, then a stabilization process occurs in which there is ample time to connect with the adjacent planar steroid-amino acid complex, and so forth until the chain of amino acids gets to the point where peptide bonds are formed resultant in a lower energy configuration, rather than positioned in a balanced configuration with a long chain of planar steroid molecules.

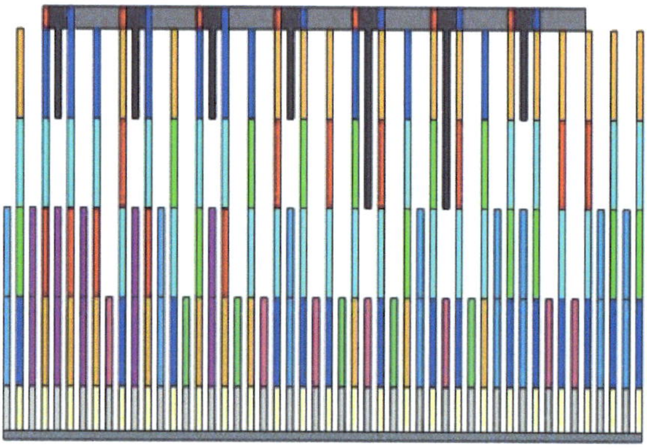

Figure 7.6: The final configuration in which the matching of the candidate amino acids within the assembly vessel is achieved. The ceiling on top is comprised of planar steroid molecules which enable an opening to evaluate candidate amino acids for stability with the interaction vessel. If unstable, for example, too big or too small, the interaction vessel moves onto the next candidate, as the candidate amino acid vacates the site. If the amino acid works with the interaction vessel, then after chain stabilization, the amino acid forms a peptide bond with the adjacent amino acid, driven by alignment of the A-T complimentary planar steroid comprising the ceiling. (In this representation, the color coding of the planar steroid molecule on the ceiling of the interaction vessel is not indicated.)

7.4 Translated DNA

To determine the amino acid sequence, the chemical and structural information of the vessel is affiliated with an amino acid that is compatible with the three-nucleotide interaction vessel. If the second base of the nucleotide triplet comprising the interaction vessel is a pyrimidine molecule, that is cytosine or thymine, the distance to the entry point is longer, and hence an amino acid with a longer residue is necessary. For the purine molecules at the base of the interaction vessel, the opposite is true, and shorter residues would be appropriate; otherwise coupling with the planar steroid molecule is not possible at the entry point. The starting point for this example is specified as the interaction vessel associated with AAA. It consists of three nucleobases of adenine, interleaved with three A-T complementary planar steroids, which make-up the walls and the docking site for the interaction vessel assembly. The first nucleotide of AAA in this example is associated with the referencing planar steroid of orientation KFBP and the nucleotide orientation of PBFK. The comparative chart of Table 6.1 indicates that structural symmetry is achieved at both walls through association with Phenylalanine (F), which is therefore the first amino acid assembled in the interaction vessel. The next nucleotide set is TAG, which has a purine molecule as the second nucleotide, and thus amino acids with shorter residues are candidates for assembly into the interaction vessel. There are also two functional groups on the wall as indicated in the structural chart. This would favor interaction with an amino acid that has a functional group; however the adenine base is apparently bound to the wall functional group, and thus a neutral molecule is enabled with methyl projections that must be consistent with the internal interactions of the vessel, including those of its walls. The choice is thus isoleucine (I). The general strategy of translating the interaction vessel construction to amino acid is indicated in the translation chart of Figure 6.6. Using this chart, the analysis is continued for the interaction vessels associated with the nucleotide sequence of AAATAGCACTGTCTCGTCGGT to identify the amino acids that occupy each of the 7 interaction vessels. After 7 amino acids are assembled, peptide bonds are formed sequentially with the planar steroid molecule of the ceiling driving the alignment and charge requirements. The resultant peptide is indicated in Figure 7.7. It is the protein structure FIVTEQP.

7.5 Recursive Table

In addition to the forward direction in which an encoded sequence is given and the nucleotides determined, there is the inverse situation to consider, that may be referred to as creation mode. In this inverse direction, that is a directed synthesis,

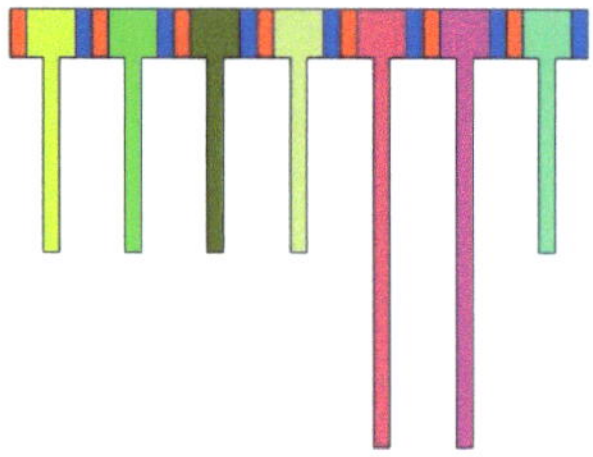

Figure 7.7: The final representation of the peptide as FIVTEQP after the elongation process is completed. The planar steroid molecular comprising the wall and the ceiling remain in place, and amino acid assembly continues until the transcription bubble closes.

the encoding sequence is also easily determined to produce the desired nucleotide sequence. In this situation, the desired nucleotide sequence or the amino acid sequence is given, and the problem is to encode it such that if the same methods were applied, it would produce the specified nucleotide sequence or the amino acid sequence. For example, in this study the first nucleotide is G, which would be either PFBK or PBFK depending on the sequence of a preceding steroid molecule, which for this problem was considered as molecule fifty, KFBP; thus the first molecule would be PFBK, since it would not have hydrogen bonding, that is would be C-G complementary, and thus there is only one choice from Table 3.2. This corresponds to the planar steroid orientation KFBP. Thus, to achieve the next nucleotide, A, would involve selecting the appropriate sequence using Table 3.2. It requires hydrogen bonding since A-T only has two internal hydrogen bonds and three are required, and A is a pyrimidine, thus it must start with P. Therefore, the choice is PBFK. This process is repeated until the encoding sequence is finished for the DNA-Steroid complex. The exact result would be produced of course since the encoding sequence is the same.

For this peptide, the interaction vessels indicated that the amino acid sequence was FIVTEQP. If only this information was provided, then there would be several choices on the nucleotide sequence. In this case, one would be selected, which would result in a nucleotide sequence. Since there is a one-to-one mapping between nucleotide and the encoding sequence, a set of interaction vessels would result that would match exactly the original protein sequence. Hence, in either case, nucleotide or protein specification, the synthesis direction is feasible. Note that this also enables a feedback loop, which implies that the resultant synthesized protein can be fed back to influence the encoded nucleotide sequence, which after processing results in a different protein, perhaps with better, or at least new, properties.

For creation mode or directed synthesis, it is possible to develop a recursive table in which the desired nucleotide sequence is given to the creator, or specified by the creator, and the objective is then to arrange the encoded starting materials such that the resultant nucleotide sequence, after the synthesis steps involved in transmission and decoding, results in the desired sequence. This contrasts to the case above in which the starting materials, that is the paired planar molecules, are randomly set in a geometrical sequence without regard to the outcome sequence; thus, after its transmission and decoding, the nucleotide sequence has an outcome, but its sequence corresponds to that which was randomly ordered at the outset. For a directed synthesis to produce a desired sequence, Figure 7.8 contains the recursive table for producing the desired nucleotide sequence. It is recursive because the reference pair is needed to specify the next nucleotide, because of intermolecular binding between the j and $j+1$ dimer. Each dimer consists of two planar steroids, denoted as $n-1$ and n. For the very first nucleotide, $j=1$, the reference configuration for $n=0$ is specified, or if circular DNA, then the last dimer element $n=2*j$ would specify the initial reference pair.

The recursive table of Figure 7.8 will now be applied to cross-check with the example nucleotide sequence `AAATAGCACTGTCTCGTCGGT` associated with the peptide FIVTEQP. The base dimer, $j=0$, was given as PFBK, KFBP. To get an `A` as the first nucleotide, the look-up table is used to find PFBK, KFBP in the first two columns (see first row). Since the nucleotide `A` is desired, the look-up table indicates that the next dimer set would be PBFK, KBFP. For the next nucleotide, the dimer set just determined is used as the referencing pair in the first two columns (see third row). To get the second nucleotide, which is also an `A`, the table is used to determine that the next dimer would be PFBK, KFBP. This procedure would then continue for the entire desired sequence. The result can be checked with the randomly generated orientation sequence listed for this example in Appendix C.

Referencing Planar Steroid Pair (j)		Desired Next Nucleotide	Next Planar Steroid Pair (j+1)	
Paired PS (n-1) To Be Nucleotide	Paired PS (n) To Be Planar Steroid		Paired PS (n+1) To Be Nucleotide	Paired PS (n+2) To Be Planar Steroid
PFBK	KFBP	A	PBFK	KBFP
		C	KBFP	PBFK
		G	PFBK	KFBP
		T	KFBP	PFBK
KFBP	PFBK	A	PFBK	KFBP
		C	KFBP	PFBK
		G	PBFK	KBFP
		T	KBFP	PBFK
PBFK	KBFP	A	PFBK	KFBP
		C	KFBP	PFBK
		G	PBFK	KBFP
		T	KBFP	PBFK
KBFK	PBFK	A	PBFK	KBFP
		C	KBFP	PBFK
		G	PFBK	KFBP
		T	KFBP	PFBK

Figure 7.8: In the creation mode of directed synthesis, the recursive table indicates the configuration of the dimer to construct the unified complex so as to arrive at the desired nucleotide sequence after transmission and decoding via the synthesis steps. As a reminder, P denotes the five-carbon ring of the planar steroid, K denotes the six-carbon ring of the planar steroid, B denotes back orientation of the ring, and F denotes forward orientation of the ring. The PS abbreviation in the table denotes planar steroid. The table can be summarized: if the next nucleotide desired is A or G, then the orientation starts with P; likewise, if C, T is desired, then the next sequence will start with K. For hydrogen bonding of a desired A or T, the rings either have to be completely opposite, or exactly the same, depending upon whether the referencing pair is a P or K; otherwise, there will be no hydrogen bonding, hence C or G.

Chapter 8

Replication

Replication is necessary for life function. The double helix form of DNA, which was formed by the simultaneous release with planar steroid molecules from a unified complex, can serve as a template from which subsequent copies can be made. Despite its use as a replication template, it is important to note that the double helix has less information content than the unified complex due to the rotation of roughly half of its base pairs, which produced the double helix configuration. This chapter will provide information on these replication topics:

- Method to produce replicated strands from the double helix using complementary planar steroid molecules as the intermediary structure.
- Applying replication techniques for repairing damage to a portion of the double helix
- Re-creating the true master image, which was associated with the unified complex that produced both structure and function.

The images produced from the double helix will contain less information than the unified complex, since the information is only of a structural capacity. The true master image, which contained both structural and functional information, will thus need to be recreated from the observable information of the double helix, if both structure and function is targeted.

8.1 Replicating the Sub-Master Double Helix DNA

The replication of the double helix will utilize planar steroid molecules as templates to guide the complementary base pair into position on the single strand DNA. After the assembly of the molecules along the single strand of DNA, and the template positioning, the in-phase growth of the phosphodiester sugar backbone will result in the replication of the double helix. Thus, with strand separation and replication of each strand, a doubling of the original material results. The planar steroid molecule is released as the template during the replication process, as it is not consumed, and is thereby available for continued replication.

In Figures 8.1(a) - (d) for A, G, C and T, the molecular model is indicated for the association of the complementary base pair with a planar steroid molecule that is configured for that particular base pair. During the transition of the unified complex into the double helix and the steroid molecule, two types were produced. One associated with the C-G pairing, which did not contain a mid-molecule hydroxyl group, and for the A-T pairing, which does have a mid-molecule hydroxyl group. Figures 8.1(e) - (h), the van der Waals spheres indicate that the planarity, spacing is appropriate for mirroring the planar steroid and deploying it as a template. Figures 8.1(i)-(l) the representation of the steroid and complementary base pair is indicated in the key format that will be used for continued analysis, showing the ring structure and functional groups comprising the steroid and base pairs, as well as its orientation. The first step for replication involves the separation into the single strand DNA. Then to prepare the strand for replication, it is filled with complementary base pairs and with the planar steroid template holding it together.

To develop the procedure for replication, an example is presented for the single strand DNA sequence of `TAGTATGCGGTTCCGACGTCTA` as indicated in Figure 8.2(a). The separation of the double helix into a single strand can take place through the binding of the steroid molecule onto the double helix, which will locally separate the strands forming a small bubble about the binding site. Additional binding will form numerous bubbles along the strands, resulting ultimately in the complete separation of the strands, some of which is driven by the incorporation of the planar steroid molecules within the section for direct transcription and translation, which can also be utilized . After filling the strands with the planar steroid molecule, the process will enable the next step of replication, as indicated in Figure 8.2(b), after association with the complementary base pair. It seems likely that replication and transcription will compete, in which transcription initially prevails, because it is locally confined process, until it is overtaken over by the broader replication process.

In Figures 8.3(a) - (d), the example is further studied to indicate the progression of the replication process after the strands are separated and the sites are filled with the planar steroid template and base pair. The filling process can occur while the replication process is proceeding. The initial step is the start of the in-phase fabrication of the phosphodiester backbone, which enables the release of the steroid

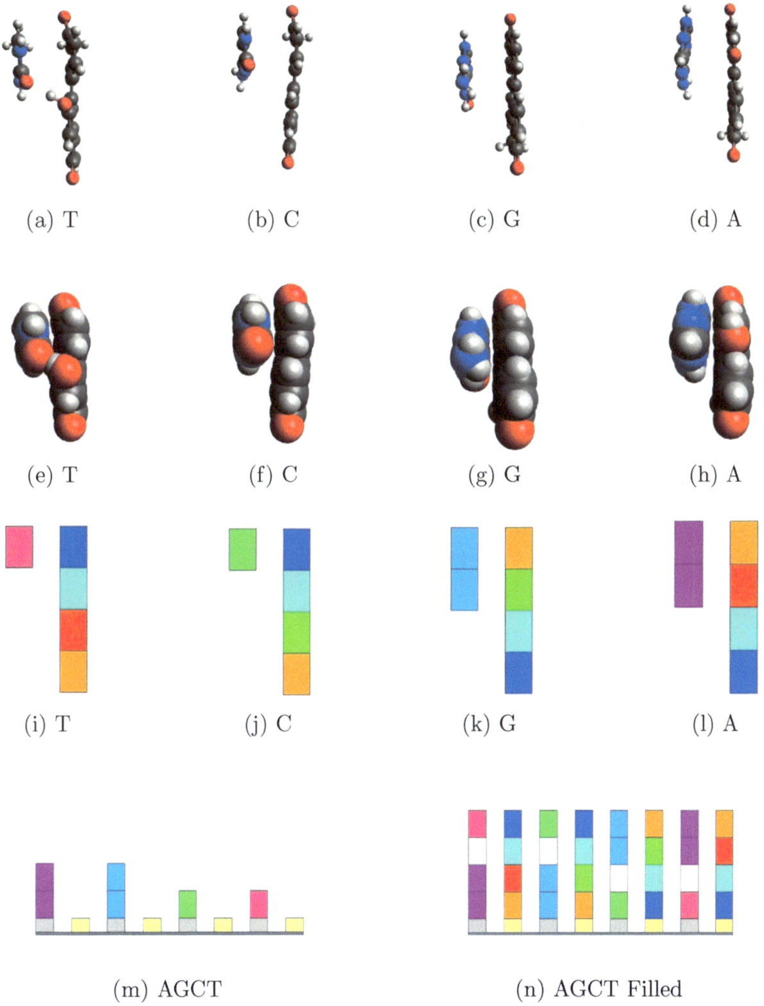

(a) T (b) C (c) G (d) A

(e) T (f) C (g) G (h) A

(i) T (j) C (k) G (l) A

(m) AGCT (n) AGCT Filled

Figure 8.1: (a)-(l) Representations of the planarity and alignment of planar steroid and complementary base pair; (m) single strand DNA; (n) ready for replication.

template as the double helix is formed, as indicated in Figure 8.3(b). This process continues until the overall process is completed. Thus, an efficient mechanism of replication is enabled through deploying the steroid molecule as a template.

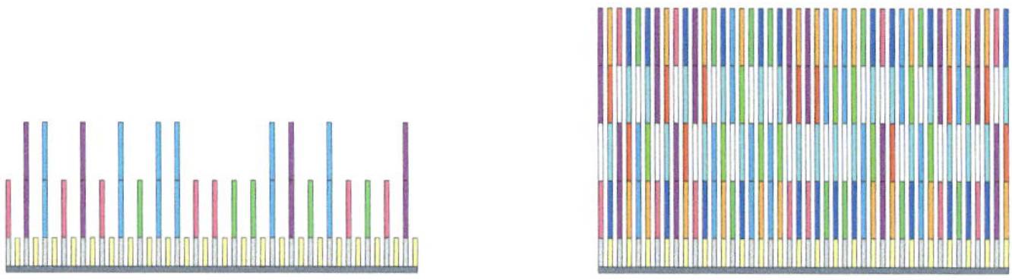

(a) (b)

Figure 8.2: (a) The example strand of DNA used in this application in replication. The sequence `TAGTATGCGGTTCCGACGTCTA` *is used. (b) After filling with the complementary base pairs and stabilizing the overall structure with planar steroid molecules.*

8.2 Damage Repair

In addition to replicating off of a single strand, the replication procedure of Figure 8.3 can be used to correct for damages in the double-stranded DNA. The planar steroid molecule together with the complementary base pair are aligned within the damaged zone, whether comprised of a single or many nucleotides using the technique presented in Figure 8.1 at the damaged zone. The phosphodiester sugar backbone is then built along the edge in the manner shown in Figure 8.3 to complete the correction procedure. This procedure is graphically indicated in Figure 8.4.

Thus, the replication procedures are also effective in the development of an error correction strategy. The solution is effective if the original code can be reconstructed, and for encoded DNA using the planar steroid configuration, it is possible to reconstruct the original encoded sequence of the orientation of the planar steroid molecules in the co-enzymatic complex, by knowledge of a fixed reference point and the nucleotide sequence. Having the planar steroid molecules available and serving as a matching template, error correction strategies are enabled. This is useful in either completely correcting for a damaged component, or at least in improving the probability that the corrected construct matches the original.

In addition, with the access to the DNA dinucleotides provided around the region defined by the binding site of the steroid molecules, error correction can be performed on the single strand since there is a template for comparison to determine if the

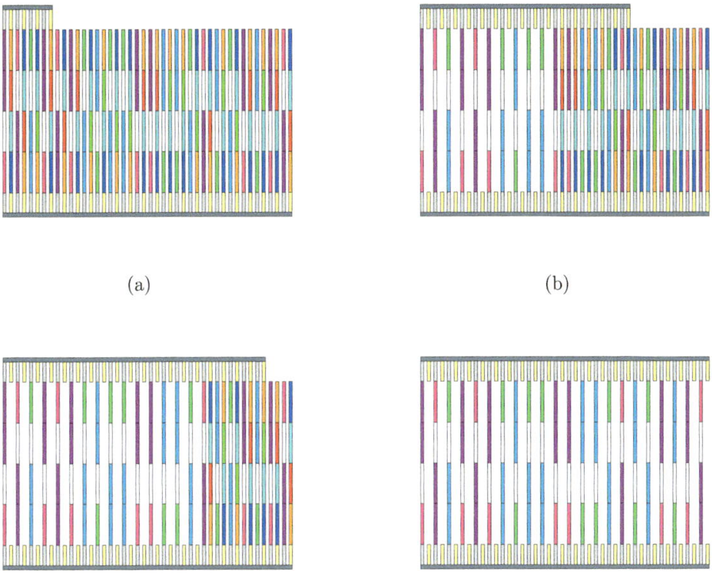

(a) (b) (c) (d)

Figure 8.3: The procedure to produce a replicated double helix DNA from a single-strand of it. Thus, a doubling in capacity from the original DNA. (a) the phosphodiester backbone begins synthesis attaching to the nucleotides; (b) after several nucleotides are formed, the DNA will be rotated into position consistent with the double helix and the planar steroids are ejected; (c) this process continues to form the double helix; (d) all the way to the end of the sequence until the phosphodiester backbone is completed.

synthesis steps were in error, or if there was a misalignment of the nucleotides. It will favor a self-correction with the planar steroid molecule providing an external force to strengthen the neighboring groups to force a realignment. Other forms of self-correction are possible, including:

- The planar steroid molecules enable a redundancy check since both the DNA

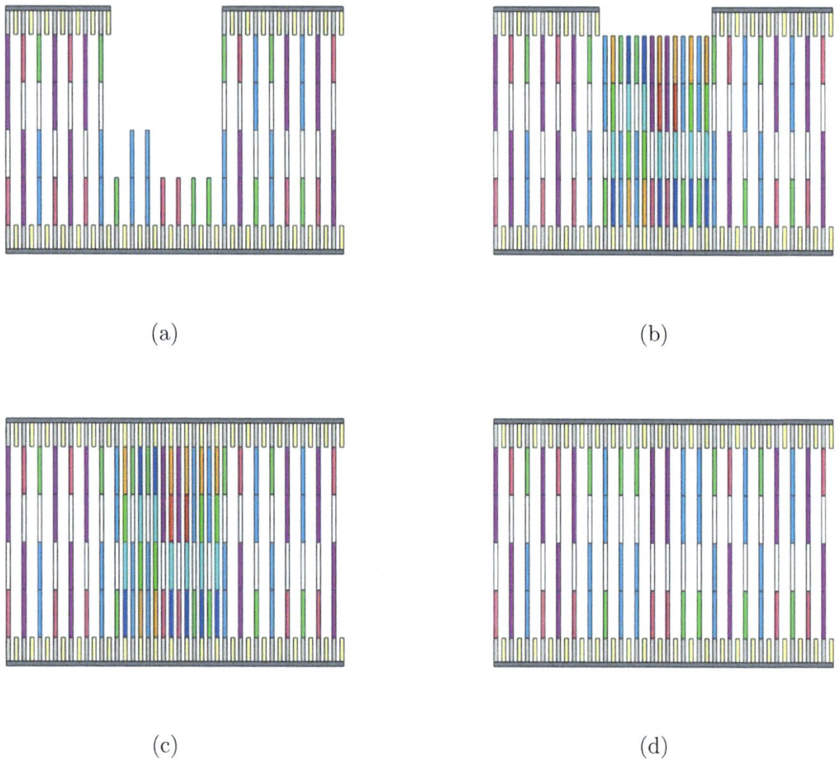

Figure 8.4: The procedure to correct for errors within the double helix is shown. (a) consider that a section of nucleotides have been destroyed and are in need of repair; (b) the planar steroid templates and complementary base pairs are attached; (c) the phosphodiester-sugar backbone is completed; (d) the double helix re-forms to eject the planar steroid molecules.

nucleotides and the planar steroid molecules were produced in a co-enzymatic way from the same molecular complex. As one influenced the other during their synthesis, the planar steroid molecule is complementary and can contribute to an error correction strategy since it is useful in recreating the original coded sequence.

- Error correction occurs through the self-organization to achieve an interleaved group of structures, in which the binding of the mid-section hydroxyl group to the thymine molecule, or when coding through its association with the non-functional group of adenine, and thus not with the amine group of guanine. When the planar steroid is not favorably placed, for example guanine is where adenine should be, there is a natural affinity to displace the planar steroid and replace it with the one that can form the hydrogen bond.

- Error correction when the DNA is not separated, before the coding begins by matching the planar steroid connection to the pair: To determine which one is incorrect, since it is possible to recreate the original code and narrow the selection as to which one is incorrect, there is improved probability of picking and correcting the appropriate nucleotide.

- Prior to translation when populating the strand, there is further error correction by the matching of the planar steroid structure to the complementary base pair. If a wall is not in place, the planar steroid molecule, which also forms the ceiling, will not be able to form an attachment point. Since the full housing is not constructed, amino acid assembly will not proceed until the structure is corrected and completed. Thus, the error will not propagate and there is sufficient time to allow for its correction.

8.3 Reproducing the True Master Image

Starting from an observable DNA sequence, any attempt to reproduce its True Master Image, which contains much more information than the derived DNA double helix, is an enormous task. The issue is that the purine and pyrimidine bases comprising the double helix are all facing the same respective direction, purines back, and pyrimidines forward. However, the true Master image, that is the unified complex, contained different orientations, which was lost when distributing structure and function, in the form of released planar steroid molecules and the DNA double helix. Thus, to reproduce the true Master image, it is necessary to recreate the original orientation from which the double helix arose. And even in this pursuit, although it is possible to determine the orientation, since the very first steroid orientation is unknown, the phase may be completely opposite from what was in the original. However, if one were to recreate the true Master image, it is possible to induce not only structure, but also function, and therefore create a critical life function.

To create the true Master image, which is unobservable, from the observable information of the double helix, there needs to be a determination of the original orientation of each base pair. Fortunately, this can be determined, and needs to be started from the first position and move in one direction. While the identification of the first nucleotide is known, its original orientation is not known, since it is defined by the proceeding steroid molecule, which is lost during the creation of a life function by releasing the double helix and the planar steroid molecules. Thus, there are two solutions to the true Master image, and they are completely out of phase, for each nucleotide. Thus, two master images would be needed, one of which will be the true master image, and the other will be its complete opposite. Both will yield the same double helix.

In Table 8.1, an example is given of the steps that are necessary to recreate the master image from the double helix. Using the original starting material, that is the first observable double helix from the true master image, the first step involves mirroring the first nucleotide, then the original steroid molecular structure is brought in place in between the first and second nucleotide, which hydrogen bonds to the first molecule, while orienting the second nucleotide for correct hydrogen bonding. Then, the first nucleotide needs to move out of the way, and the first and second original starting material need to move upwards, to allow a third molecule of the starting material to mirror the newly oriented second nucleotide. This process then repeats for the entire chain. The result is a new unified complex of the True Master Image. The phase will be in accordance with the starting phase of nucleotide one.

Master	Nucleotide	Observable	**Step 1** Mirror	**Step 2** Insert	**Step 3** Remove	**Step 4** Mirror	**Step 5** Insert	**Step 6** Remove	**Step 7** Mirror	**Step 8** Insert	**Step 9** Remove
PBFK			KBFP	KBFP	KBFP	KBFP	KBFP	KBFP	KBFP	KBFP	
KBFP	C	KB	KB	KB	PBFK	PBFK	PBFK	PBFK	PBFK	PBFK	KBFP
PBFK				PBFK		KFBP	KFBP	KFBP	KFBP	KFBP	PBFK
KBFP	T	KB	KB	KF	KF	KF	KF	PFBK	PFBK	PFBK	KBFP
PFBK						KBFP	PFBK		PBFK	PBFK	PFBK
PBFK	G	PF	PF	PF	PF	PF	PB	PB	PB	PB	PBFK
KBFP										KBFP	KBFP

Table 8.1: This is the procedure to produce the True Master Image of the Unified Complex based only on the observable nucleotide sequence of the single strand of DNA. It involves the mirroring of the nucleotide by the original material comprising the True Master Image, followed by the insertion of a planar steroid molecule of the original material. Through hydrogen bonding, the next nucleotide is properly oriented during the insertion process. The base pair on the single strand is then able to shift away from the complex, removed. These three steps are continued until the replication process is complete.

8.4 Reproduction Hierarchy

The hierarchy of reproduction is consistent with the notion that from a unified complex, both structure and function were produced for life function. After life started, it was only necessary to reproduce structure, since function was available through the planar steroid molecule. This implies that there was enough planar steroid molecules from which functional operations of transcription, translation, and replication could go forward. In less abstract terms, structure was the double helix and function was associated with the planar steroid molecules. The planar steroid molecules can be used to directly reproduce from a single strand of the double helix without any loss of information content associated with the double helix. However, there is a loss of information of the unified complex, which contained encoded information on the planar steroid molecules as functional construction devices. But since the planar steroid molecules are generated, their further usage would imply enough material to accomplish its task in transcription, translation, and replication. From the process flow analysis, there is a sufficient quantity of material available as these steps do not require the external addition of planar steroid molecules.

Chapter 9

Genetic Rearrangement

To express genetic information in a diverse manner, several types of splicing strategies are available for the configuration of planar steroid molecules and the DNA double helix. This occurs to increase the complexity of the protein derived, as well as confer optimal properties in subsequent replication steps. The types of splicing strategies for the planar steroid-DNA complex utilizing replication techniques, include: (1) differential splicing of genetic constructs to generate a variety of proteins; (2) rearrangement of expressions (exons) during replication by removal of non-expression areas and obtaining a larger, combined expression. Other areas of interaction, such as the defect generation misplacement of an expression inserted into a non-expressive area, or in the genetic rearrangement by recombination would also appear feasible under certain conditions.

9.1 Splicing Expressions

To generate a variety of proteins from the same strand, differential splicing is a mechanism in which expressions connect during the translation process to create longer chains of proteins and arranged in different segments. While the mechanisms to produce alternative splicing schemes are known with modern-day mRNA, in this section, it is shown that it is also possible to perform differential splicing, connecting series of expressions, through the use of the planar steroid and a single strand of DNA.

An example for the sequence `TGTACATGATCTCCGACGGGTACAAAATCGCG` is represented in Figure 9.1, which consists of the expression 1 `TACATGATC` and expression 2 `TACAAAATC` that contain a start and stop nucleotide triplet and an amino acid, `F` for expression 1 and `Y` for expression 2. In this case, the splice is shown for expression

1 to expression2, which would occur by either the physical distance reduction by motion of the strand between the two sections that are covered by planar steroids, bypassing the non-expression section that does not contain a planar steroid ceiling, attaching `F` of expression 1 to `M` of expression 2, to result in the protein `MFMY`.

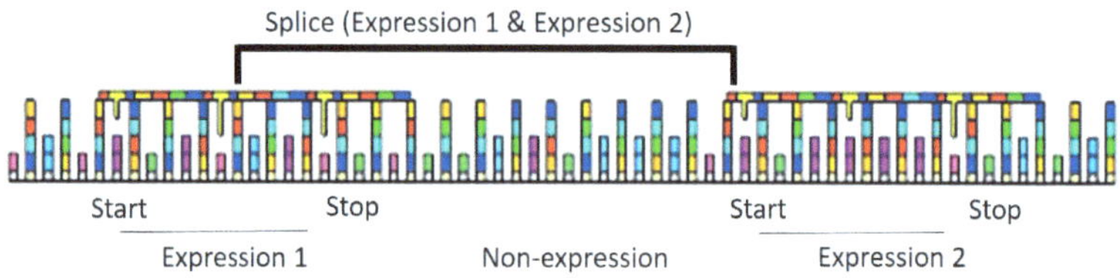

Figure 9.1: Splicing two expressions together using the planar steroid molecule as the intermediary. In areas where the start and stop mechanics are not available, it is possible to splice the expressions together. The Stop site is immobile due to the strong interaction of the amino acid (whose structure can only be assumed since it is non-expressive) with the interaction vessel. Thus, the last expressed amino acid is removed from its position and interacts with the Start amino acid of the next expression. This forms a longer protein structure, and increases the variety of proteins which can be generated.

9.2 Removing Non-Expressions from DNA Strands

During the replication process, it is possible to remove non-expressions, and thus deploy only expressions in the replicated strand. Thus, greater lengths of proteins can be built. This can happen if replication takes place concurrently with translation, in which a splicing event is taking place. If it occurs significantly afterwards, unless the strands are well stabilized connecting the two expressions, the opportunity for bridging will be lost, at least on this phase of replication. Figure 9.2(a) and (b) indicates this procedure. Two separate strands of DNA will result in the next replication cycle.

9.3 Insertion and Recombinant DNA

There are other possibilities using this strategy for rearranging the genetic composition via planar steroid molecules. This includes expression (exon) insertion in an existing expression, which can occur during the replication phase. In addition, during

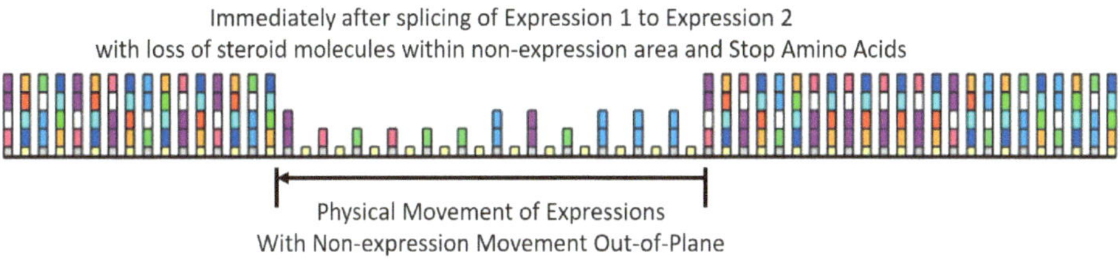

(a) Immediately after Transcription

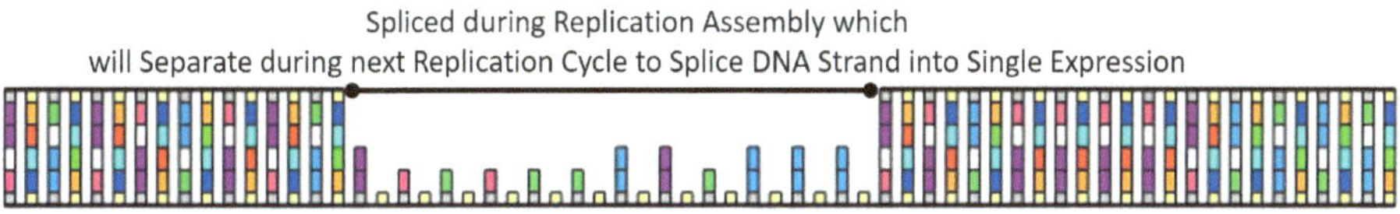

(b) Skipping non-expression during Replication

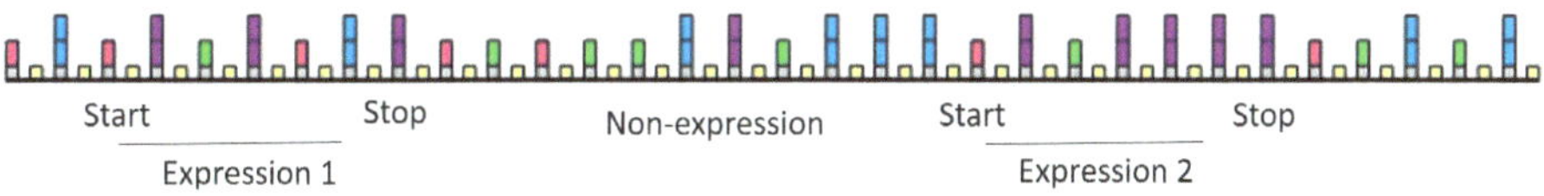

(c) Strand 1: After Next Replication Cycle

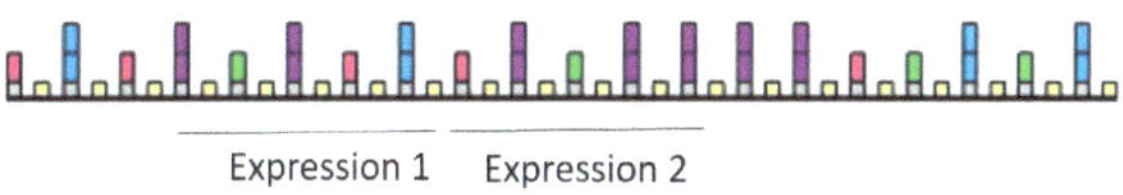

(d) Strand 2: After Next Replication Cycle

Figure 9.2: (a) immediately after splicing expression 1 to expression 2, and if the planar steroid molecules are detached from the non-expression due to the motion of expression 2 towards expression 1, there is an opportunity to combine expression 1 and expression 2 into a single expression during the replication cycle. (b) If the timing is fast enough, the two expressions can become spliced together during the replication cycle, of which the single strand DNA will separate during the next replication cycle. The original DNA containing both expressions separated by a non-expression will remain in place (c), and the single strand of combined expression without the non-expression will result in (d).

the replication phase, genetic recombination of DNA can occur. Both rearrangement processes occur by the strands coming in proximity to the replicating strand, and the transition from one part of the strand to the other during the phosphodiester backbone procedure guided by the planar steroid templates. A spacing model for recombinant DNA is shown by arranging planar steroid models in various crystals to show that the replication direction can be altered when implemented in conjunction with nucleotide base pairs.

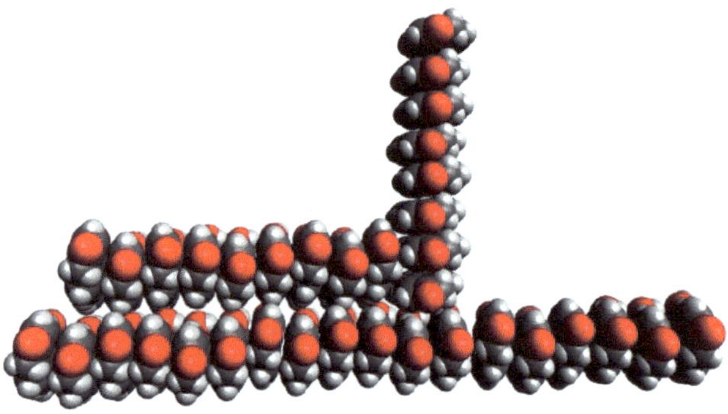

(a) Intersected

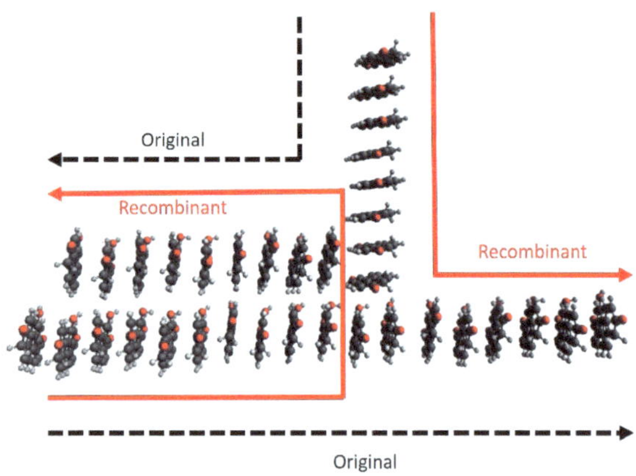

(b) Recombined

Figure 9.3: A spacing model to indicate feasibility of recombinant DNA formed via jumping onto an intersecting strand as the phosphodiester backbone is processed .

Chapter 10

Ensemble Encoding Patterns

It is practical to analyze the alternative design patterns of the planar steroid molecule, which ultimately, with in-phase binding, comprise the DNA double helix, as well as the complementary planar steroid set to enable function. In a realistic environment, there will be contamination to the perfectly pure set of planar steroid dimers arranged with hydrogen bonding at the ends and internal alignment. A realistic set of ensembles of planar steroid molecules will include dimers whose internal rings are not aligned, but mismatched, and there will be random inserts of planar steroid molecules into the stacked set, either as a randomly arranged dimer, or as a single unit, randomly inset into the chain. It is thus useful to determine the impact that this has on the capability to form DNA sequences at base four. In this section, such realistic dimer sets are considered, including the possibility for intra-dimeric bonding configurations and completely randomized complexes. The ensemble effect of aligned, unaligned, random, inter-dimeric, and intra-dimeric coupling is discussed and evaluated. This chapter considers the phosphodiester binding to be in-phase, that is covalent bonding on each end of the planar steroid, which is a pre-requisite for forming DNA. The chapter after this one will consider out-of-phase binding configurations, which result in the phospholipid bilayer enclosing DNA. Thus, the alternative binding configurations provide a richer array of constructs from which to develop genetic materials.

10.1 Unaligned Internal Rings

For the inner rings, alignement corresponds to a setting for example of (PFBK,KFBP), whereas an unaligned arrangement of the inner rings would be (PFBK,KBFP). The comparison of aligned and unaligned planar steroid molecules is considered in Figure

10.1. In Figure 10.1(a) and (b), the two sides of the dimer of planar steroid molecules is indicated in the standard arrangement with the internal two rings aligned, as shown in this electrostatic potential map. The result indicates a balance of five hydrogen atoms and four hydrogen atoms on both sides, a total of nine hydrogen atoms in between the oxygen groups on both sides of the dimer, which are engaged in hydrogen bonding. However, in Figure 10.1(c) and (d), which has the inner two rings unaligned, there are ten hydrogen atoms on one side, that is five atoms and five atoms, and on the other side there are only eight atoms, four atoms and four atoms. This imbalance will cause less planarity in matching the two planar steroid molecules, whereas with alignment there is better balance.

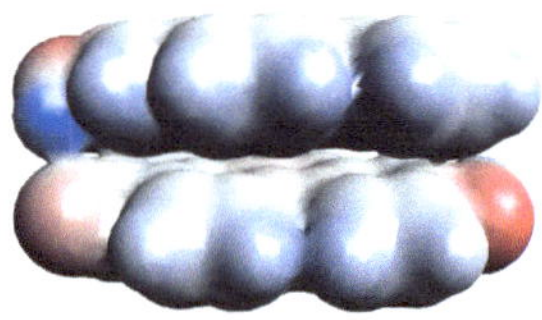

(a) Aligned

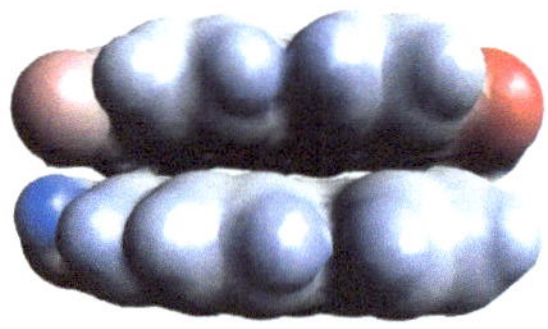

(b) Aligned

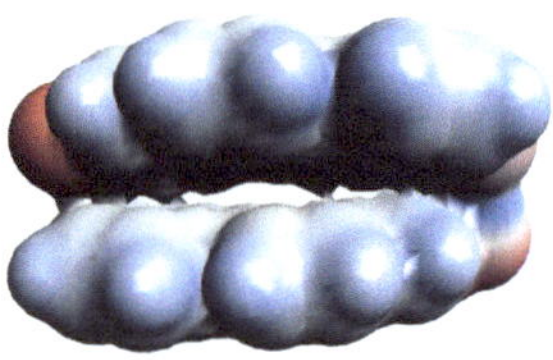

(c) Unaligned

(d) Unaligned

Figure 10.1: In (a) and (b), the two sides of the dimer of planar steroid molecules with the internal two rings aligned, indicating a balance of five hydrogen atoms and four hydrogen atoms on both sides, a total of nine hydrogen atoms in between the oxygen groups on both sides of the dimer, as indicated in this electrostatic potential map. However, in (c) and (d), which has the inner two rings unaligned, there are ten hydrogen atoms on one side, that is five atoms and five atoms, and on the other side there are only eight atoms, four atoms and four atoms. This imbalance will cause less planarity than for the case in which the number of hydrogen atoms are equal on each side through internal alignment.

Even with unaligned dimers, it is still possible to generate a recursive table to design the nucleotide sequence according to a pre-planned form, as indicated in Figure 10.2. As with the aligned dimers, the interdimeric bonding configuration determines the outcome of the nucleotide and planar steroid sequence. The primary change is that the inner two rings of the prior and subsequent orientations to the interdimeric coupling are unaligned.

Referencing Planar Steroid Pair (j)		Desired Next Nucleotide	Next Planar Steroid Pair (j+1)	
Unaligned PS (n-1) To Be Nucleotide	Paired PS (n) To Be Planar Steroid		Paired PS (n+1) To Be Nucleotide	Unaligned PS (n+2) To Be Planar Steroid
PBFK	KFBP	A	PBFK	KFBP
		C	KBFP	PFBK
		G	PFBK	KBFP
		T	KFBP	PBFK
KBFP	PFBK	A	PFBK	KBFP
		C	KFBP	PBFK
		G	PBFK	KFBP
		T	KBFP	PFBP
PFBK	KBFP	A	PFBK	KBFP
		C	KFBP	PBFK
		G	PBFK	KFBP
		T	KBFP	PFBK
KFBP	PBFK	A	PBFK	KFBP
		C	KBFP	PFBK
		G	PFBK	KBFP
		T	KFBP	PBFK

Figure 10.2: For the case of unaligned inner two rings, a recursive table that is similar to the one for an aligned inner two rings can be generated. P denotes the five-carbon ring of the planar steroid, K denotes the six-carbon ring of the planar steroid, B denotes back orientation of an inner ring, and F denotes forward orientation of the inner ring. PS denotes planar steroid. As in the case for alignment, the table is summarized the same as in Figure 7.8 in which if the next nucleotide desired is a purine, A or G, then the orientation starts with P; likewise, if a pyrimidine, C or T, is desired, then the next sequence will start with K. For hydrogen bonding of a desired A or T, the rings either have to be completely opposite, or exactly the same, depending upon whether the referencing pair is a P or K; otherwise, there will be no hydrogen bonding, hence C or G.

10.2 Intra-Dimeric Coupling

It is interesting to evaluate the situation in which the binding pattern involves an association of an intermolecular co-enzymatic process associated with the dimer itself, and not with the interdimeric coupling. In this situation, either an aligned pair of dimers will be co-processed independent of the next pair, or an unaligned pair, will be processed independently of the next dimer. By analyzing the resultant nucleotides, insights on the ensemble process can be assessed.

In the case of the aligned dimer, the sequence will be among the four: (PFBK, KFBP), (PBFK,KBFP), (KFBP, PFBK), or (KBFP, PBFP). Using the tables presented in Chapter 3, the result, for example considering the sequence as the nucleotide, the result for four will be C, C, G, G. Thus, the sequences will consist of only C or G nucleotides from the 5' strand and will be base two. This situation is observable as CG islands.

For the opposite case in which the dimers are unaligned and the co-enzymatic complex takes place at the dimer, instead at the inter-dimeric pairing, the sequence will be among the four: (PFBK, KBFP), (PBFK,KFBP), (KFBP, PBFK), or (KBFP, PFBP). Thus, using the tables presented in Chapter 3, the result, for example considering the sequence as the nucleotide, the result for four will be T, T, A, A. Thus, the sequences will consist of only T or A nucleotides from the 5' strand and also will be base two. This arrangement would result in AT islands.

10.3 Randomized Complex

For randomized arrangement of planar steroid molecules, in which the complementary hydrogen bonding at the ends is not utilized, and rather if the molecules are randomly brought together as single units or as dimers, either aligned or unaligned, without particular preference, it is still possible to develop a recursive table to determine the nucleotide sequence from the specified chain. The only piece of information would be that the binding is in-phase, and it needs to be specified which alternating set of planar steroids are bound to the phosphodiester chain. In Figure 10.3, the recursive table is presented for a random chain of planar steroids. The specification of the nucleotide may be performed at a sequence of two, and then the planar steroid at the next in the sequence is random. In the table E denotes the end group, and is randomly drawn from the {P, K} while O is the orientation of the inner two rings is drawn randomly from the set {F, B}. $\bar{E}$ denotes the complement of E, and $\bar{O}$ denotes the complement of O.

Referencing Planar Steroid Pair (j)		Desired Next Nucleotide	Next Planar Steroid Pair (j+1)	
Random PS (n-1) To Be Nucleotide	Paired PS (n) To Be Planar Steroid		Paired PS (n+1) To Be Nucleotide	Random PS (n+2) To Be Planar Steroid
EOŌĒ	KFBP	A	PBFK	EOŌĒ
		C	KBFP	EOŌĒ
		G	PFBK	EOŌĒ
		T	KFBP	EOŌĒ
	PFBK	A	PFBK	EOŌĒ
		C	KFBP	EOŌĒ
		G	PBFK	EOŌĒ
		T	KBFP	EOŌĒ
	KBFP	A	PFBK	EOŌĒ
		C	KFBP	EOŌĒ
		G	PBFK	EOŌĒ
		T	KBFP	EOŌĒ
	PBFK	A	PBFK	EOŌĒ
		C	KBFP	EOŌĒ
		G	PFBK	EOŌĒ
		T	KFBP	EOŌĒ

Figure 10.3: For the case of randomized rings, a recursive table that is used to design, or its inverse can be used to determine, the next nucleotide is generated in which the activity of the molecular coupling at the co-enzymatic site determines which nucleotides and complementary planar steroid results. The nomenclature is P denotes the five-carbon ring of the planar steroid, K denotes the six-carbon ring of the planar steroid, B denotes back orientation of an inner ring, and F denotes forward orientation of the inner ring. PS denotes planar steroid. E denotes the end group, and is randomly drawn from the {P, K} while O is the orientation of the inner two rings is drawn randomly from the set {F, B}. Ē denotes the complement of E, and Ō denotes the complement of O.

10.4 Ensemble Encoding

In a real situation there would likely be a range of patterns available from the planar steroid molecules: aligned dimers, unaligned dimers, and single planar steroids of random orientation with respect to the growing chain. Thus, the binding patterns for the co-reactive site that defines hydrogen bonding capability would be comprised of inter-dimeric and intra-dimeric couplings. Depending upon the relative concentrations of the groupings, the resultant nucleotide sequence would have a range of CG islands, AT islands, and a nucleotide sequence of a random and structured components. However, the yield would be an equivalent number of nucleotide pairs and complementary planar steroid molecules. Thus, independent of the concentrations, the sequence will be put into action since the planar steroid molecules are available for immediate function.

An example is used to consider the case of ensemble encoding. Consider the example presented in Chapter 7 with the ring orientation presented in Appendix C, which was shown to correspond to the sequence `GAAA TAGCACTGTCTCGTCGGTTGG`. If a random planar steroid molecule of KFBP was inserted after molecule 6 and after molecule 18, the new sequence would be `GAA TCGCCGCC TGTCTCGTCGGTTGG`. Thus, there is a transformation of the leading and trailing nucleotide due to the inserts, and a conversion of six nucleotides to a CG island of `CGCCGC` which replaced `TAGCAC`.

Chapter 11

Molecular Primer of the Genetic Cell

As in the case of the molecular primer for DNA, which showed the trace of the steroid molecule existed in all nucleotides, the phospholipid bilayer of the Cell also contains a similar molecular primer. Further, this molecular primer is integrative with DNA and the Unified Complex, and as will be seen, results in the molecular Cell membrane and enclosed genetic components to be generated at the same site using the same starting material, that is the planar steroid molecule. In this section, a molecular primer is developed for the phospholipid bilayer, which is compared with DNA and the planar steroid molecules that comprise the formative stage of the encoded Unified Complex.

11.1 Primer of the Phospholipid Bilayer Construct

In Figure 11.1(a), the overall construct of the phospholipid is indicated. It shows a linear arrangement of carbon molecules, of which the length is typically shown as 17, but according to references can range between 15 and 21. In the diagram of Figure 11.1(b), of particular interest is the phosphor grouping. The primers or clues to the origin of the phospholipid bilayer comprises three structures: (1) the phosphate group with three carbon elements on one side of the ester, and the three elements, two carbons and one nitrogen, on the other ester; (2) the carbon tail comprising approximately 17 carbon atoms arranged as a linear chain; (3) the overall structure of the phosphate head on both sides of the phospholipid bilayer, which is indicated in Figure 11.1(c). The dimension of the phospholipid is 19.8 Å of the 17-carbon linear chain, and 5.0 Å distance between the two linear chains.

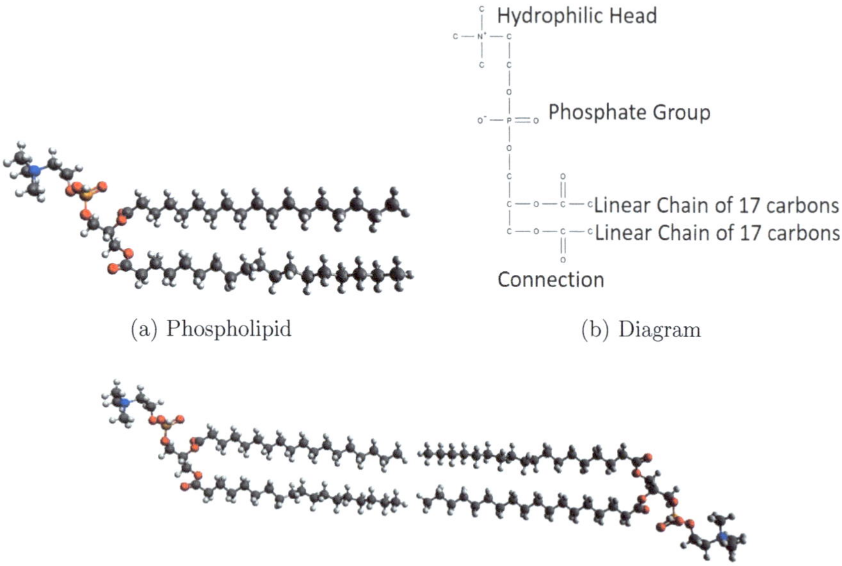

(a) Phospholipid

(b) Diagram

(c) Bilayer

Figure 11.1: Schematic of the phospholipid bilayer, which will be analyzed in this chapter to formulate a primer for the Cell, similar in concept to that used for DNA. (a) The phospholipid comprises a phosphorous connection to hold onto a hydrophilic head and hydrophobic tail. (b) The diagram shows that there is a connection of a carboxylic acid group to a linear chain of carbons, which is 17 carbons long, but can range from 15 to 21, according to the literature. (c) The bilayer format is shown, in which the tails face to form a hydrophobic core with hydrophilic heads on either side.

11.2 Correlation to DNA Nucleotide Paring

To relate these results to identifying the origin and its correlation with genetic material, an examination of the DNA nucleotide is indicated in Figure 11.2, which shows similarity with the phospholipid bilayer as follows: (1) there are three carbons on either side of each phosphate group; (2) there is a gap of approximately 15 - 20 Å between the two phosphate groups; (3) the phosphate groups face each other, separated by the molecular primer of the complementary base pairs comprising the DNA molecule. This information will be used in the following chapter as the basis for indicating the method and mechanism of forming the molecular genetic cell, which will contain the DNA material as well as the planar steroid molecules.

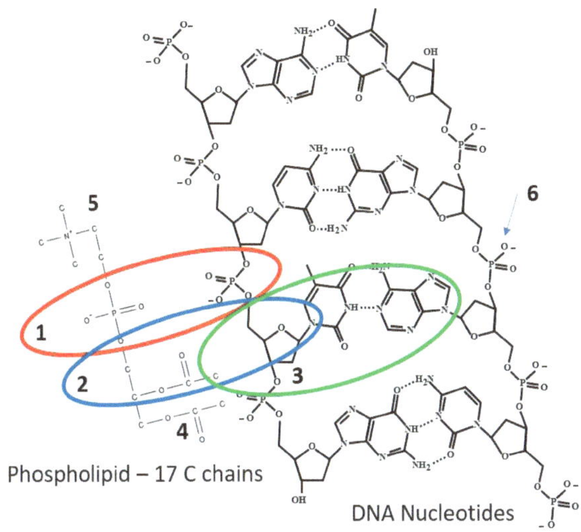

Figure 11.2: As a primer, there is a correlation of the phospholipid molecule to the DNA nucleotide. This will serve as a guide in the subsequent chapter to correlate the unified complex to the biological cell. There are several areas of relationships, which includes the phosphodiester backbone correlated to the effective backbone comprising the phospholipid. In addition the linear chain of 17 carbons is the same number as the steroid complex, which was shown in Chapter 2, on mapping to the DNA nucleotide pair. The connector is like the sugar molecule. The phospholipid bilayer, with phosphate groups on either side, is like the phosphate groups on either side of the DNA molecule. The hydrophilic head is similar in the number of elements on either side of the phosphate group.

To summarize the correlation with the DNA nucleotide pairs, Table 11.1 lists the areas of comparison. These include structural and chemical relations. The structural equivalency includes the 17-carbon chain which is equal to the number of carbons comprising the planar steroid studied in the previous chapters on the design of DNA nucleotides. This indicates that there is a correlation with the unified complex which was used to derive the double helix. Now the attention turns to the phospholipid bilayer of the biological cell.

In yet another structural relation of the molecular primer comprising the DNA nucleotides to that of the phospholipid bilayer, it is noted that cholesterol molecule

Area	Primer
1	Phosphophate is one to one comparison between lipid and DNA
2	Connector comparison to sugar
3	A linear chain of 17 carbons that comprise the intiating steroid molecule of the unified complex
4	the axial spacing is consistent with the steroid molecule adjacent spacing
5	three carbons of DNA are equivalent to two carbons and the attaching R group
6	Phosphate group on the other side of the DNA molecule is equivalent to the second layer from the phospholipid bilayer

Table 11.1: A list is presented of the areas of comparison where the structural and chemical characteristics of the phospholipid bilayer correspond to the DNA nucleotide, and its molecular primer, the planar steroid molecule. This will enable a relationship with the biological cell, DNA, planar steroid molecules, and the structure of the unified complex.

as indicated in Figure 11.3 penetrates the membrane of biological cells. Thus, there is compatibility of the steroidal structure with the phospholipid bilayer. It also indicates that there is proximity of the molecule to the phospholipid bilayer. This information will be used in the following chapter.

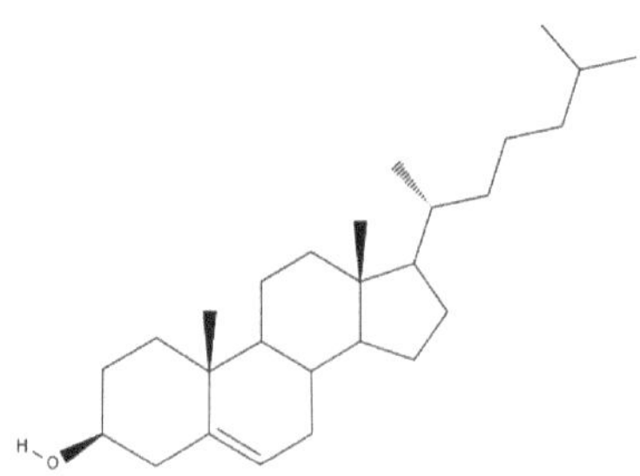

Figure 11.3: Molecular structure of cholesterol also shows four fused rings, of which cholesterol is found in phospholipid bilayers. This is seen in living cells and is indicative of compatibility of ring structures with the phospholipid bilayer, and thus is consistent with the developments here of a primer for the Cell.

Chapter 12

Design of the Genetic Cell

This origin of the cell is described in this chapter using the clues described in the previous chapter, as well as the clues described in the construction of the Unified Complex, which was presented in Chapter 3.

12.1 Origination of the Phospholipid Bilayer from the Unified Complex

Figure 12.1(a) indicates a portion of the unified complex after it is encoded, and prior to its attachment to the phosphodiester backbone. In attaching the phosphodiester backbone, which is the first set of steps, prior to the enzymatic transfer of one steroid nucleotide to a paired set of complementary bases, there is a 50% chance that the phosphodiester attaches in-phase as the ends are secured, which means that both sides of the alternating nucleotides are covalently attached to the same nucleotide. This in-phase attachment is indicated in Figure 12.1(b). This configuration will proceed in the previously described manner to result in the double helix and planar steroid molecules, which will enable its function. However, the out-of-phase attachment is indicated in Figure 12.1(c). In this case, the reaction to transform the components to DNA nucleotide pairs does not take place as the enzymatic complex is offset by one phosphate group. Thus the strands ultimately move away from the other, until there is large separation distance, as shown in Figure12.1(d). This results in a phospholipid bilayer comprising the cell enclosure. Hydrogen bonding in the center will enable this form of the phospholipid bilayer to stabilize since the hydroxyl group of the six-carbon ring will interact with the ketone group of the five-carbon ring.

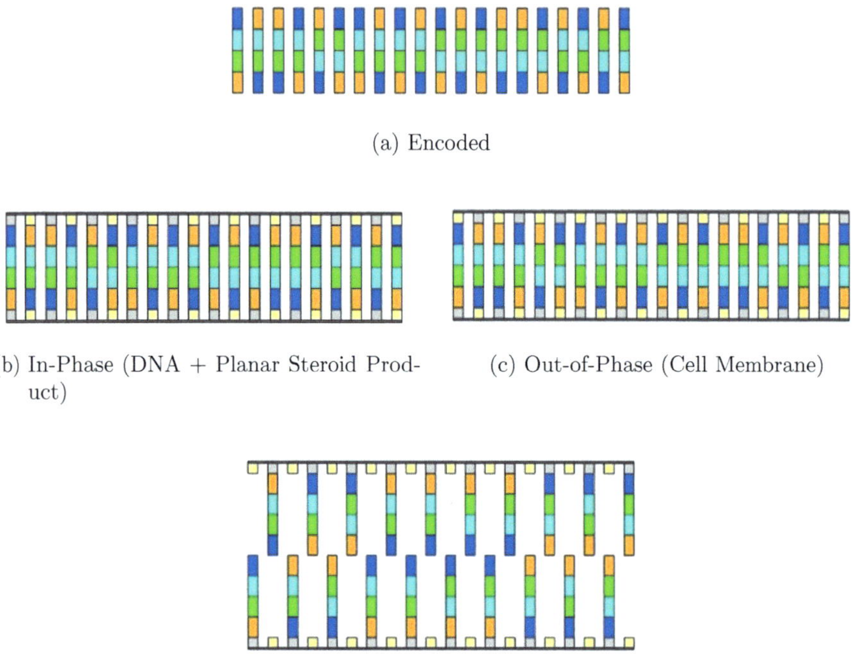

(a) Encoded

(b) In-Phase (DNA + Planar Steroid Product)

(c) Out-of-Phase (Cell Membrane)

(d) Out-of-Phase Separation

Figure 12.1: These figures represent a fundamental derivation of the origin of DNA and the biological cell, all from the same starting material and assembly. (a) The encoded structure is comprised of a set of planar steroids with orientations that define the nucleotide sequence as a base four construct. (b) In-phase coupling of the phosphodiester backbone to the encoded structure, in which in-phase denotes the covalent bonding on both sides of the same molecular structure (gray), and hydrogen bonding to the phosphate group (yellow). If it is in phase, it will result in the DNA double helix and complementary planar steroids after synthesis steps. (c) The out-of-phase representation, which happens when the covalent bonding in forming the phosphodiester chain is mismatched, strand to strand. (d) If it is out-of-phase, there is separation of the strands, as the phosphodiester chain pulls the covalently bound steroid molecules apart. This structure will become the phospholipid bilayer of the biological cell after synthesis steps within the same environment as that responsible for DNA and the genetic code.

The molecular model of the linked phosphate chain is indicated in Figure 12.2(a). This structure has the similar arrangement of structures to permit the function of the modern-day phospholipid bilayer. There is a hydrophilic head on either side, and a hydrophobic tail in the center. The distance between the two hydrophilic heads is roughly 30 Å, which is within the range of the modern-day phospholipid bilayer. Through the mechanism described in the Appendix, the rearrangement and separation of the groups are indicated in Figure 12.2(b), which would give the modern-day version of the cell. In Figure 12.2(c), a van der Waals representation indicates the hydrophobic and hydrophilic representation of the steroid molecules bound to the phosphodiester backbone. Note that the hydroxyl group of the planar steroid molecule undergoes hydrogen bonding to its counterpart on the other side of the phosphodiester chain. Also, the results are applied to the 17-carbon structure of the planar steroid molecule, but it is noted that to cover the full range of a phospholipid bilayer length from 15-21 carbon elements, the results are extendible to include an additional ring, hence 21 carbon elements, or a break in one of the internal rings, hence 15 carbon elements, which would imply irregularities in the formation of the dimeric pairs.

12.2 The Molecular Cell and its Genetic Content

The spherical structure of the molecular cell and the feasibility for the circular nature of the DNA, especially in lower forms of life, is also consistent with the structure of the unified complex. In Figure 12.3(a), thc circular structure of the encoded structure is indicated. The phosphodiester backbone represented as an in phase construction is indicated in Figure 12.3(b). (It is noted these developments allow for both linear and circular DNA.)

The result of the double helix is indicated in Figure 12.3(c) for the DNA circular structure of:

$$\texttt{DNA}_{\texttt{circular}} = \texttt{GACTGGTCAAGTTTTGCTATGCCTAGTTTCTGAGATCGACATACTGAACG} \quad (12.1)$$

In forming a cell through an invagination process of linear bilayer constructs, multiple genetic DNA components can become entrapped within the cellular structure unable to escape. Thus, a variety of different expressed genomic content, in the form of circular and linear genetic components could reside within a cellular construct. This completed structure with multiple levels of genomic content within a cell is indicated in Figure 12.4(a) for the encoded complex of a potential 100 nucleotide,

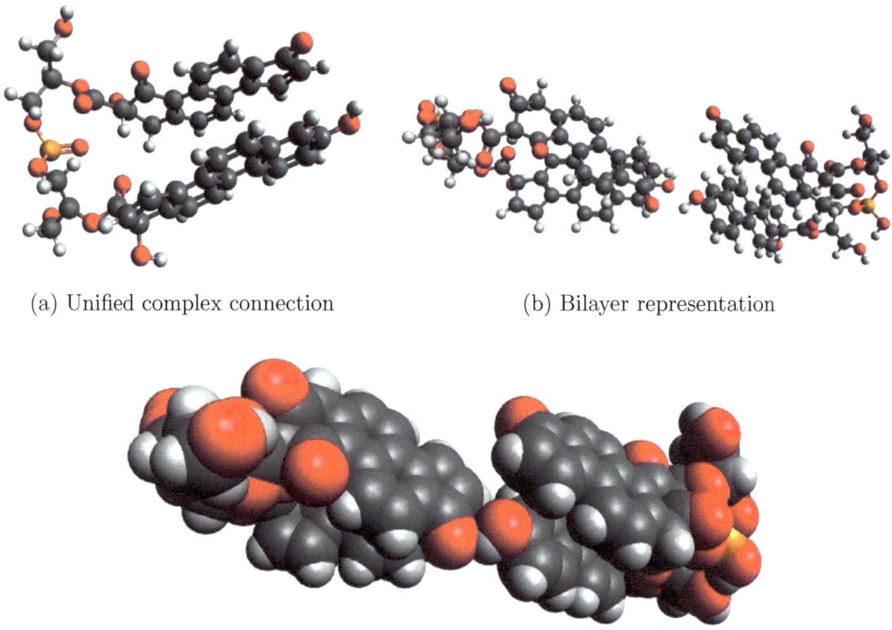

(a) Unified complex connection (b) Bilayer representation

(c) van der Waals representation

Figure 12.2: (a) molecular model of the connection of the planar steroid complex to the phosphodiester backbone in an out of phase situation where the planar steroid molecule pulls away from the other side of the unified complex. It is noted that the planar steroid molecule has seventeen carbon elements. (b) A bilayer representation indicating a largely hydrophobic interior and a hydrophilic head. (c) The van der Waals spheres indicates the hydrophobic and hydrophilic nature of the complex.

100 steroid complex; thus initially there are 200 planar steroid molecules connected to the phosphodiester backbone, except as an out-of-phase representation of Figure 12.4(b), in which one side of the steroid molecule is covalently bound, and the other side only undergoes a hydrogen bound. Thus, there will be a pull-out of the structure, and the phospholipid bilayer pre-cursor is formed, as indicated in Figure 12.4(c).

To form an enclosed structure would involve additional out-of-phase rings assembling together, driven by hydrophilic and hydrophobic forces, as shown in Figure 12.5 in which two encoded structures form an enclosed structure, with the exception that

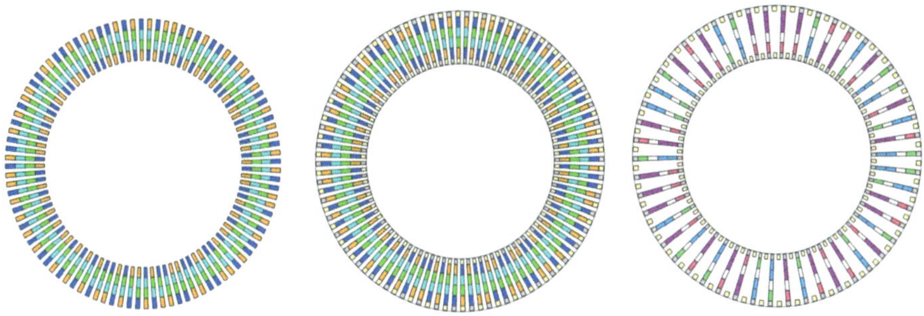

(a) Encoded (b) In-Phase (c) Double Helix

Figure 12.3: An example of circular DNA from an encoded unified complex. (a) the encoded structure in which the last steroid molecule aligns with the first nucleotide. (b) An in-phase binding of the phosphodiester backbone, in which covalent bonds are formed on both sides of the steroid molecules, which will become nucleotides, interleaved with planar steroid molecules that are hydrogen bonded to the phosphate group. (c) This will result in a circular DNA structure using the methods previously described.

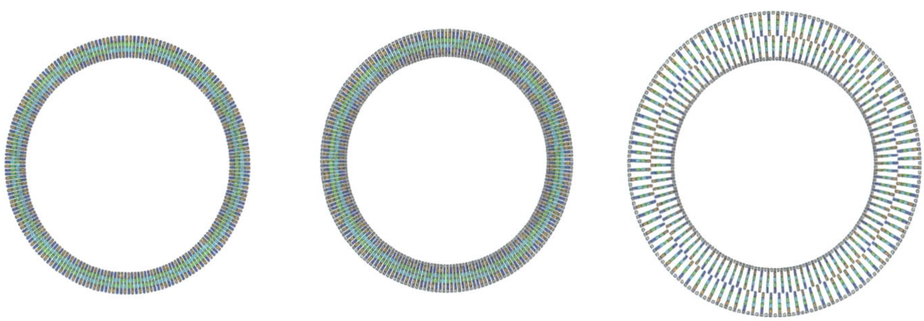

(a) Encoded (b) Out-of-Phase (c) Phospholipid Cell Membrane

Figure 12.4: (a) 200 planar steroid molecules arranged in an encoded representation. (b) out-of-phase connection of the phosphodiester backbone, in which one side is covalently bonded and the other side is hydrogen bonded. (c) the separation of the layers according to the covalent bonding configuration, thereby resulting in a phospholipid bilayer, and the biological cell structure in two dimensions.

the inner structure is in-phase connected, but the outer structure is out-of-phase. The spherical structure can also form from linear sheets of bilayer membranes involving the inward folding of the phospholipid bilayer comprised of planar steroid molecules.

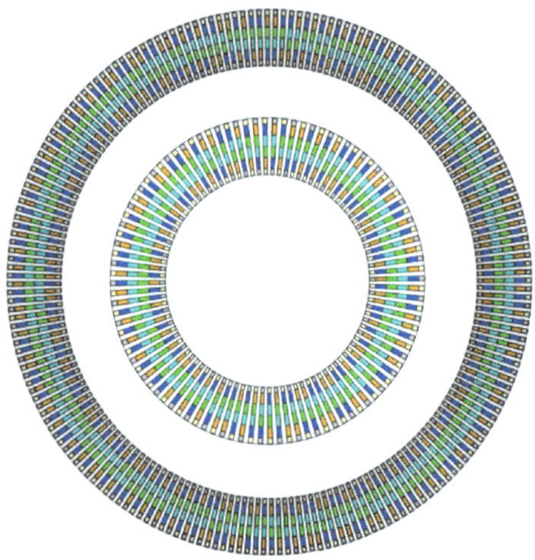

Figure 12.5: In forming the genetic encapsulated cell, there are multiple layers in which the cellular structures are incorporated within the same environment. This depicts the situation prior to pull-out to form the phospholipid bilayer, and prior to the formation of the unified complex, release of the planar steroid molecules, and formation of the DNA double helix.

Thus, after synthesis steps, there will result an enclosed double helix DNA molecule inside a phospholipid bilayer as indicated in Figure 12.6. In addition, the steroid molecules that were ejected during the formation of the double helix will be contained within the cell and the ability to immediately begin transcription is enabled.

To produce proteins, there is the need for a high concentration of each amino acid. The synthesis environment involves the arrangement of the phospholipid bilayer, produced from an out-of-phase unified complex prior to attachment from the amine group, and the double helix, formed from the unified complex with complete synthesis via an in-phase complex. Thus, there is the potential for a partially synthesized out-of-phase unified complex to become the source of amino acids. It is interesting to

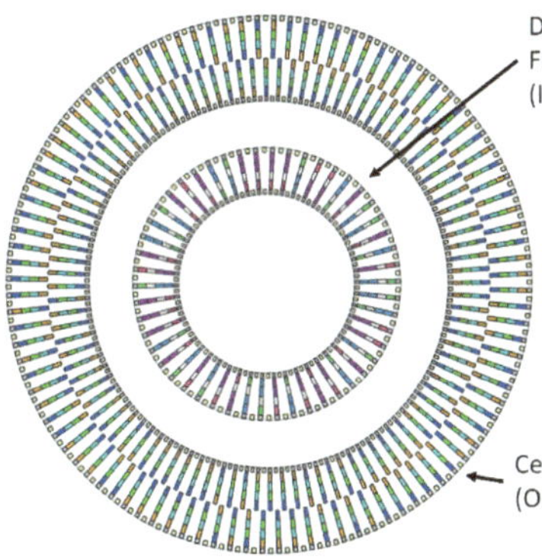

Figure 12.6: Out-of-phase binding generates a phospholipid bilayer comprising the cell enclosure for DNA. It is a spherical structure, and here the cross-section is shown to enclose the DNA double helix, which is generated through in-phase binding of the unified complex from the same starting material. The DNA double helix is depicted in circular format, although it can also be linear. In addition, the cell enclosure contains the planar steroid molecules necessary to carry-out transcriptional, translational, and replication function. For peptide generation, preliminary indications are that the amino acid set is derivable from an out-of-phase binding configuration from the same starting material and assembly of planar steroid molecules, followed by subsequent synthesis steps. Thus, the basic materials needed for the origination of genetic constructs are available in sufficient concentration within an enclosed environment to permit life function.

evaluate the point at which the pull-out from the unified complex due to the out-of-phase configuration of the unified complex binding to the phosphodiester backbone. With the amino acid configuration, which is initiated in the attachment phase, there are numerous configurations for attachment to the alpha carbon position of the amino acid.

In Figure 12.7, the starting point for the amino acid synthesis is the coupling point of the phosphodiester backbone to the unified complex. However, with an out-of-phase configuration and with some partial synthesis before pull-out and further reaction, therefore it will not have the capability to produce the phospholipid bilayer, but rather the amino acids. The configuration of the coupling point would have efficiently generated the amino acids in the proper orientation as well.

Through reaction mechanisms, it is feasible to configure a reaction sequence to

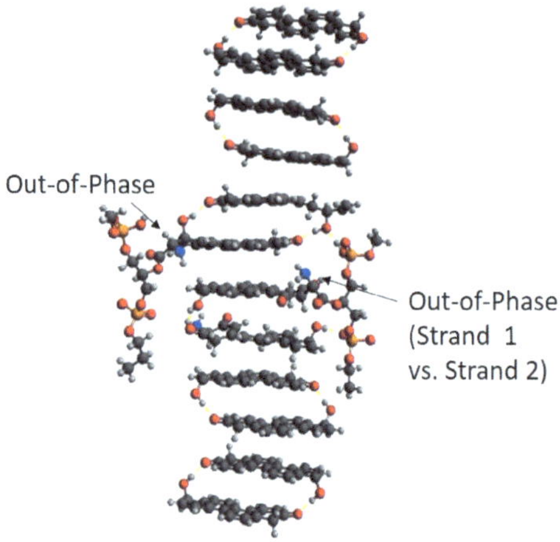

Figure 12.7: An initiation point for the manufacture of amino acids originate from an out-of-phase, partially synthesized unified complex. The starting point for the amino acid processing steps is the coupling point between the unified complex and the phosphodiester backbone.

produce the chain of carbons, oxygen, nitrogen, as well as the sulfur components that could have relevance within the interaction vessel. This would have enabled the assembly of the amino acids to form proteins, and thus would have conserved all elements. The sulfur element is the only one that would have to gain entry to the vessel outside of the range of elements that comprise the unified complex.

The synthesis mechanisms for each of the twenty amino acids, as well as the stop signals, are under development as of this writing, but in Table 12.1, the starting point for the synthesis as either the purine side or the pyrimidine side of Figure 12.8 is indicated. These mappings of the attached planar steroid complexes in a partial processed state for each amino acid is described in the Appendix A.1.6. The stop signal will emanate from both the purine and pyrimidine structures.

	Pyrimidine	**Purine**	**Either**
Amino Acids	F, Y, Q, N, D, E	S, C, W, L, P, H, I, M, T	R, K, V, A, G

Table 12.1: Starting point for amino acid production are associated with the attachment points of the pyrimidine or purine side from an out-of-phase connection.

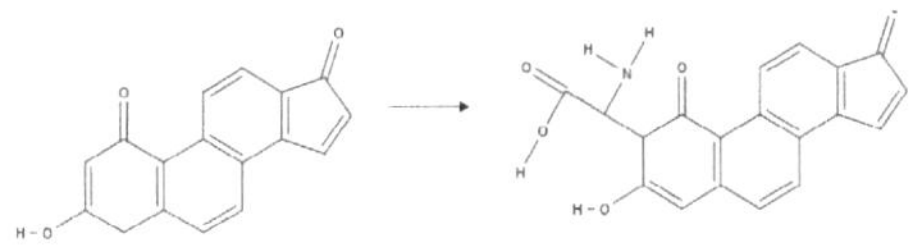

(a) Pyrimidine Start

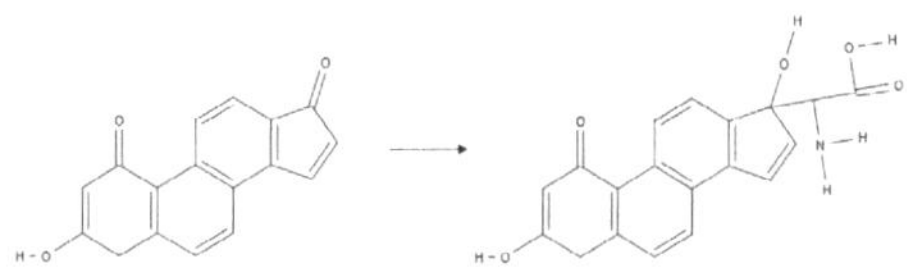

(b) Purine Start

Figure 12.8: The starting point for the amino acid synthesis from the out-of-phase representation can be the carbon of the pyrimidine side (a) or the purine side (b) of the planar steroid molecule partially synthesized towards nucleobases.

12.3 Supporting Cellular Materials

As with the out-of-phase binding producing the phospholipid bilayer, with immediate pull-out of the strands, or with amino-acid production involving partial synthesis, there is another source of cellular materials useful for genetic purposes that arise from full synthesis prior to separation of the strands. These components comprising a single-strand of planar steroid matter useful for inclusion bodies involving hydrophobic core, that would be useful for intercellular and intracellular transport mechanisms, nucleobases useful for replication and repair, and AMP, GMP, CMP, TMP that are also useful for replication and repair.

In Figure 12.9, the out-of-phase binding components in which full synthesis is achieved prior to strand separation is indicated. Note the hydrogen bonding in place between the intermolecular planar steroid and thymine, as well as between adenine and thymine. The components for this assembly include dAMP and thymine. The dAMP will be arranged in a linear chain and decomposed into single units through degradation processes. The planar steroids are arranged on a long strand and can form as inclusion bodies and for transmembrane proteins since a hydrophobic core will be formed with a hydrophilic shell and exterior.

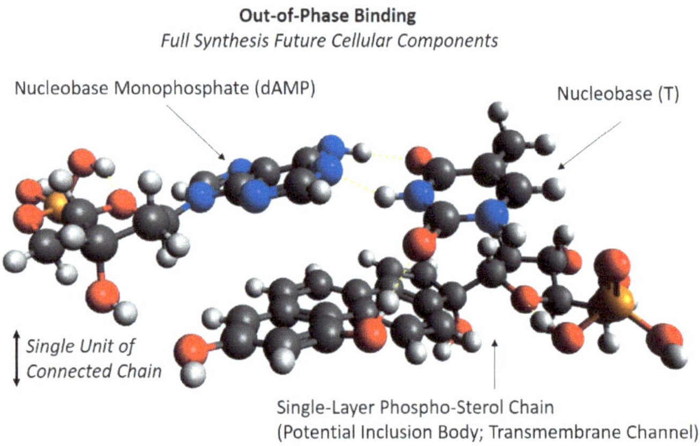

Figure 12.9: For out-of-phase binding and full synthesis progression until there is a separation of the strands, a range of additional cellular components results. This includes the adenosine monophosphates and nucleobases. Also, the planar steroid molecules connected onto the linear phosphodiester chain are available to self-assemble into components that contain a hydrophobic core and hydrophilic exterior, for example for use as inclusion bodies and transmembrane channels.

12.4 Cellular Life Function

With the availability of amino acid configurations, which has sufficient concentration within the enclosed cellular environment, all the necessary components are present for protein synthesis within one place in suitable concentrations. The overall design for the phospholipid bilayer, the double helix, the steroid molecules, and the amino acid set are all derived from a single starting material, the planar steroid. Thus, from the single unified complex, all the materials are available to generate life forms.

In Figure 12.10, the new findings are put into context. As indicated in the first submission, DNA and the genetic code are produced through the stacking of a planar steroid in a dimeric configuration, with an interdimeric co-catalytic reaction sequence with in-phase binding at both strands of a phosphodiester linkage. As the binding and organization associated with the genetic code is not of probability one, there is the need to evaluate alternative binding and dimer orientation scenarios. In addition to the binding pattern resultant in DNA, there are other useful binding pattern configurations and planar steroid arrangements that produce biologically relevant structures which improve the viability of DNA and life function. In sum, depending upon the binding arrangements, the new finding indicates that from the same planar steroid molecule and synthesis pathway, a living cell results with sufficient genetic material to succeed in growth and replication.

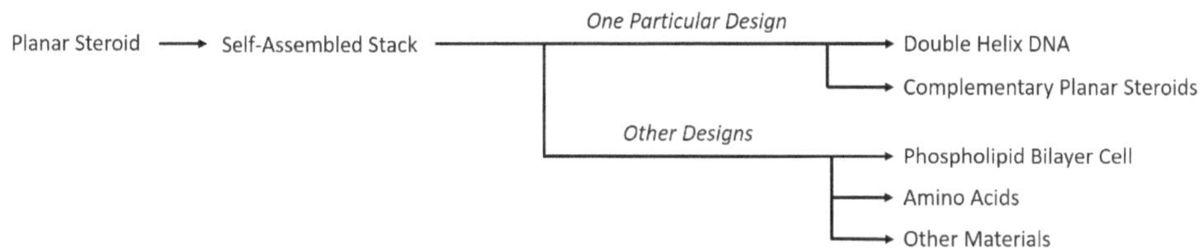

Figure 12.10: From the same starting material, components necessary for the design of DNA, its cell enclosure, as well as the life functions of transcription, replication, and translation are enabled, located in the same place, at the same time in sufficient concentrations.

To summarize the processes for cellular life function, the overall diagram is depicted in Figure 12.11. A source of planar steroid molecules self-stacks for realistic conditions into an ensemble comprised of dimers with the inner two rings aligned, which is favorable, but also there will be unaligned dimers, and randomized single planar steroid molecule, which are independent of a dimer configuration. As the self-assembly process is forming chains of the planar steroid molecules, the phosphodiester linkage is established along the ends of the planar steroid molecules, achieving covalent binding alternating with hydrogen bonding, at both ends. If the covalent bonds are attached at either end of the same molecule, then the binding pattern is in-phase. After synthesis, the result is the DNA double helix and a complementary set of planar steroid molecules that enables genetic functions and growth processes such as transcription, replication, and translation. If the covalent bonds are not on the same molecule, then a covalent bond will be attached only at one end, with hydrogen bonding at the other end. This configuration is said to be out-of-phase. In this context, there is the situation of strand separation. If the separation occurs without any co-enzymatic synthesis, then the result is the precursor to the phospholipid bilayer comprising the cellular enclosure, and intracellular membranes. If there is partial synthesis and then separation, ultimately the result is an amino acid set useful for further processing. If there is full synthesis, then the result is nucleobases, nucleobase sugar monophosphate complexes, and a single strand of planar steroid molecules. This strand can collapse on itself to form a hydrophobic core and a hydrophilic exterior, which can be used as inclusionary bodies, transmembrane channels, and other components.

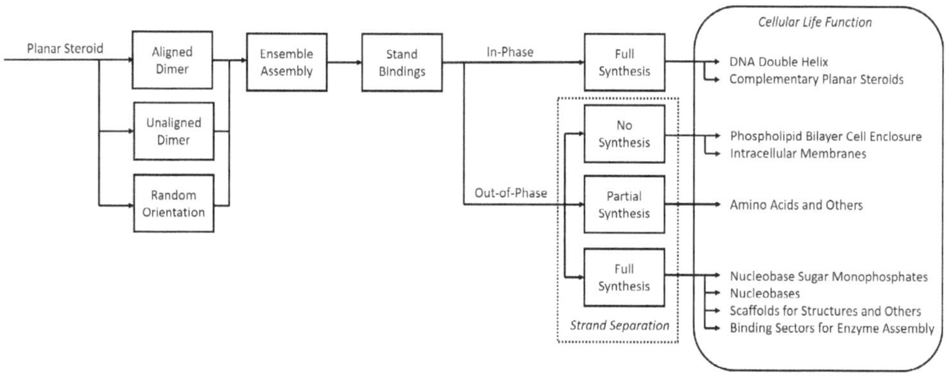

Figure 12.11: The overall process diagram is shown for the development of cellular life function from a single set of planar steroid molecules, that is life origination. A source of planar steroid molecules self-stacks into an ensemble comprised of dimers with the inner two rings aligned, which is favorable, but also as unaligned dimers, and as a random single planar steroid molecule independent of a dimer. With self-assembly of the planar steroid molecules, the phosphodiester linkage is established along the ends, either at the hydroxyl group on the six-carbon ring or the ketone group at the five-carbon ring at the other end. Depending upon the relative binding arrangement between the two strands, the in-phase configuration is destined for the DNA double helix and complementary set of planar steroids, while the out-of-phase is destined for several routes. If the strands separate rapidly, without synthesis, the result is the phospholipid bilayer comprising a cellular enclosure. If there is partial synthesis, the result is a set of amino acids. If there is full synthesis, the result are nucleobases, nucleobase sugar monophosphates, and planar steroid molecules arranged along a single strand that can become units with a hydrophobic core and hydrophilic exterior.

Chapter 13

Material Balances and Process Flows

The material balances and process flows are examined for the origination of DNA and its subsequent use for transcription. In Figure 13.1, the starting material of the planar steroid is indicated, and compared against the products of the planar steroids after the DNA double helix has been synthesized. There will be a reduction of half the planar steroids that started the process, as half will be used in synthesizing the DNA product. The number of nucleotides formed will be equal to the number of planar steroids available, thus it is not necessary to add additional planar steroids. Further, the number of A-T complementary planar steroids are equal to the number of A-T nucleotides, and there will be an equal number of C-G complementary planar steroids.

The process flow is described by use of an example in Figure 13.2 for the production of DNA and its planar steroids that will enable the DNA functions of replication, transcription and translation. To discuss quantities of materials, the formation of 50 dimers, which consists of 100 planar steroids, is evaluated in this example. The synthesis strategy produces a unified complex, requiring that the phosphodiester linkage is in-phase, and the binding pairing is the facing planar steroids of interdimeric structures. The unified complex in this example consists of 50 DNA nucleotide pairs interleaved with 50 planar steroids, as the synthesis strategy involves using an adjacent planar steroid molecule to encode the nucleotide. Because the orientation of all purine or pyrimidine off of the 5' strand are not guaranteed to be in a uniform direction, and are more likely to be distributed fifty percent in the opposite direction to that present in the double helix, thus of the 50 dinucleotides, roughly 25 will be oriented in the opposite direction from the standard, that is purines in a back direction and pyrimidines in a forward direction, relative to the four ring structure of the planar steroid. In this example, further, we consider that 30 A-T pairs were formed,

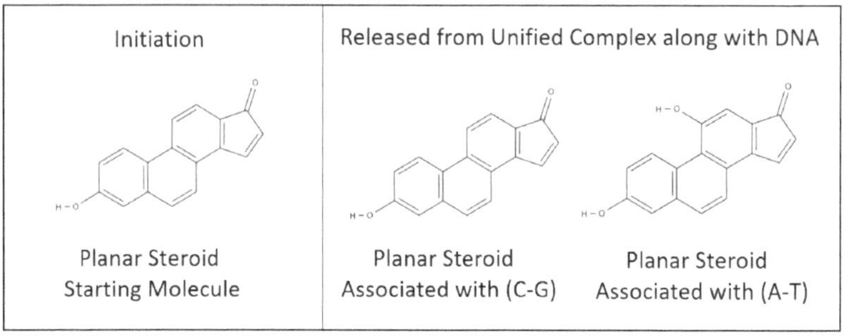

Figure 13.1: The starting material and the resultant planar steroid products. The starting material is capable of stacking and bonding end to end, while forming a symmetric structure in the inner two rings. The final products are destined for interaction with either the C-G dinucleotide or the A-T dinucleotide. The hydroxyl group for the A-T complementary planar steroid is perfectly positioned to bond to thymine of the A-T dinucleotide, thus forming three hydrogen bonds to match that of C-G. Since C-G already has three hydrogen bonds, the C-G complementary planar steroid does not require a hydroxyl group on the C-ring.

and 20 C-G pairs were formed. Thus, there will be 30 A-T complementary planar steroids, and 20 C-G complementary planar steroids. After the synthesis steps are complete, the instability of the unified complex will result in the rotation of the 25 unoriented nucleotides into their standard position to match the 25 already in position, resulting in a 50 nucleotide double helix DNA, with 30 A-T planar steroids (i.e. with a C-ring hydroxyl group) and 20 C-G planar steroids. These can then be used for the transcription and replication process.

The planar steroids utilized in the transcription process is indicated in Figure 13.3. The procedure begins with the use of planar steroids to induce strand opening by binding upstream of the transcription site, for example by use of three or more A-T planar steroids (which means that the planar steroids have a hydroxyl group that enables complementary hydrogen bonding with thymine). This will effectively knot the strands, stabilizing it at a point, but inducing instability at the surrounding sites. With the transcription bubble opened, through a self-assembly stacking process, the planar steroids are aligned with the corresponding nucleotide. In this example, 14 nucleotides are exposed along a single strand, 8 A-T and 6 C-G, thus there will be associated 8 A-T planar steroids, and 6 C-G planar steroids that will align to each nucleotide. For transcription, there are four possible codon openings, and thus four amino acids and four A-T planar steroids, which have a hydroxyl group to bind onto the third codon post, are used to form the ceiling. This results in a four amino

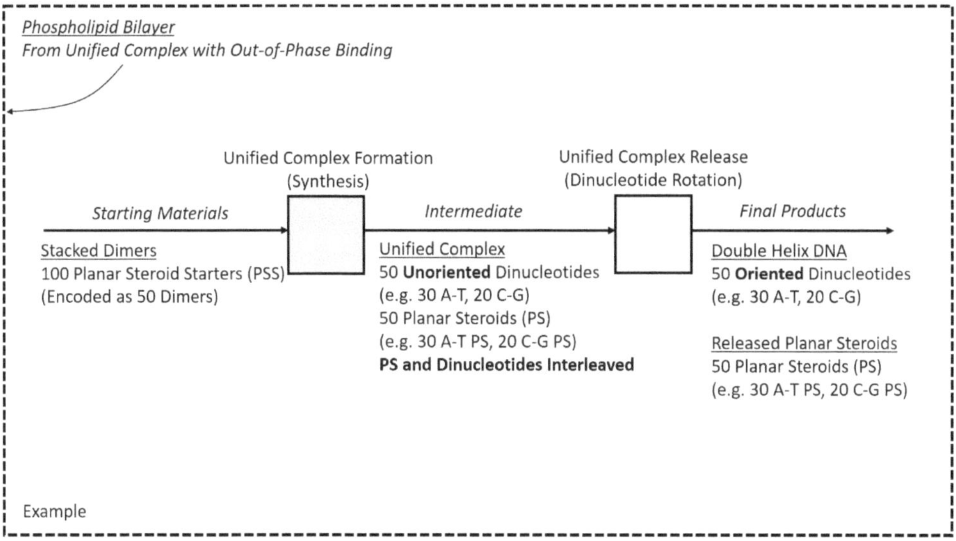

Figure 13.2: The process flow to produce DNA and planar steroid products targeted for A-T and C-G interaction. The starting materials are also planar steroid molecules that form dimers. When the dimers stack, depending upon their orientation and binding pattern, a base 4 code is produced. This assembly is stabilized by a phosphodiester chain, and further synthesis of the covalently bond molecules, using the adjacent steroid as a catalytic site. To form the DNA double helix, driven by entropy considerations, the dinucleotides that are not oriented in the manner consistent with the double helix, rotate and eject the planar steroids. The planar steroids are complementary to the dinucleotides and undergo future processing with the DNA double helix.

acid polypeptide. The entire process takes place inside the bilayer cell membrane, which results from an out-of-phase binding of the unified complex. The amino acids come from the out-of-phase binding product, in conjunction with partial synthesis and degradation of the unified complex.

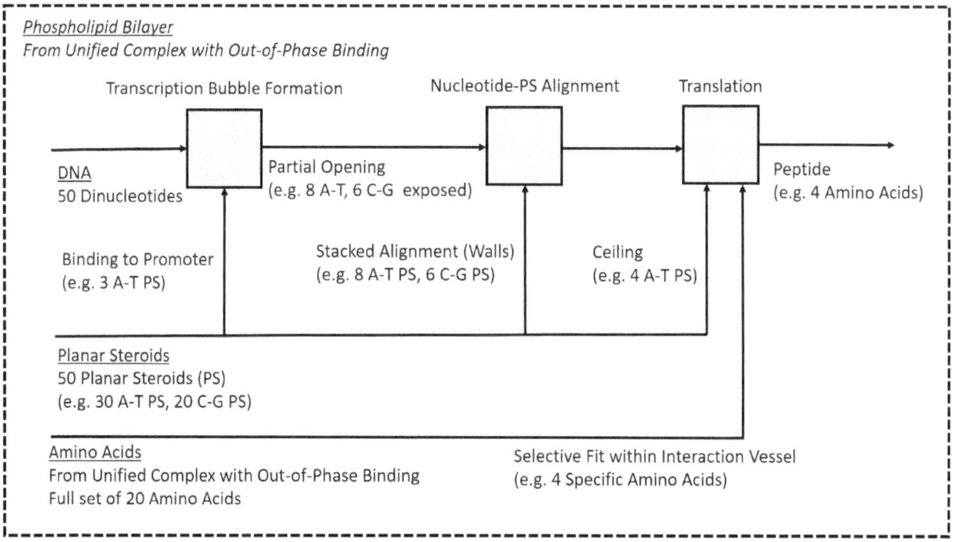

Figure 13.3: The procedures of transcription and translation to produce a protein. After DNA and the complementary planar steroid products are produced, a certain fraction of planar steroids associates with the DNA double helix, opening-up the strands by a binding process to create a transcription bubble. The planar steroids then associate through a stacking self-assembly process with an open strand of DNA to form a series of interaction vessels for assembly of amino acids at every third nucleotide. The amino acids are then linked by peptide bonds of which the planar steroid molecule comprising the ceiling provides alignment. The figure includes material balances on the planar steroids to indicate that external material is not required for transcription and translation. It can all be performed within the cellular environment formed from out-of-phase binding of a planar steroid self-assembled complex.

Chapter 14

Evolution of Genetic Constructs from Planar Steroid Assemblies and Binding Patterns

Now that the process that resulted in the origin of life is known as described in the proceeding chapters, it is interesting to compare the basic molecules and processes used to initiate life function with those in modern-day use. In this chapter, the steroid hormones are evaluated first, which is appropriate since the starting material for life was the planar steroid molecule. Moreover, it is of the form that drove this research, that is the molecular primer of DNA. Next, the evolution to the central dogma of molecular biology from the life's origination, the planar steroid assembly, is described, in which specialized binding patterns and rearrangement patterns are utilized to indicate a method and means to form tRNA, mRNA, and rRNA. Therefore, it is shown that it is feasible to start from the planar steroid molecule and produce the molecular machinery of living cells, which thereby provides additional evidence that the planar steroid was the formative molecule responsible for the genetic code.

14.1 Molecular Structures to Initiate Transcription

The planar steroids used in the developments on the origin of the genetic code are compared in Figure 14.1 with its modern-day steroid hormone counterparts of cortisol-like and testosterone-like. It is seen that the steroid molecules used to initiate life contain double bonds throughout the structure, whereas this feature is lost in the modern-day steroid hormones of living cells. Furthermore, in the modern-day version of steroid hormones in living cells, methyl groups have evolved within the steroid molecular structure in critical areas where the hydroxyl group binds onto thymine.

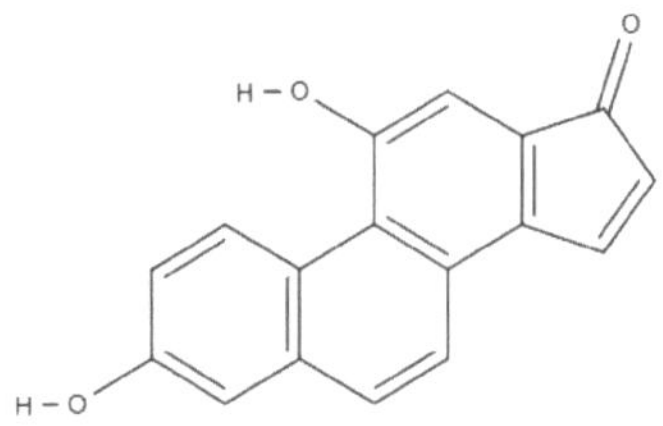

(a) Planar Steroid A-T Complementary

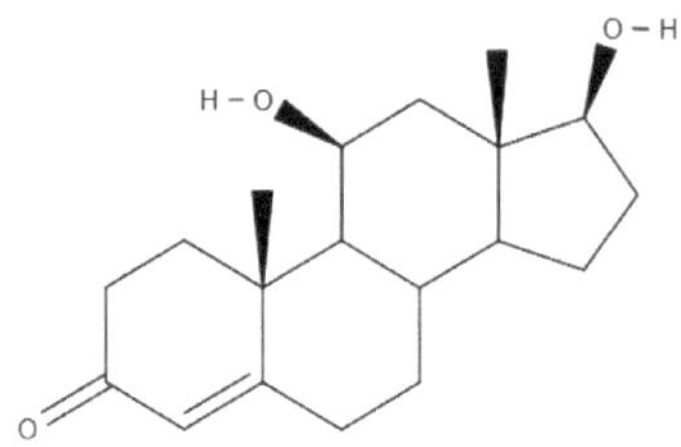

(b) modern-day Cortisol-Like

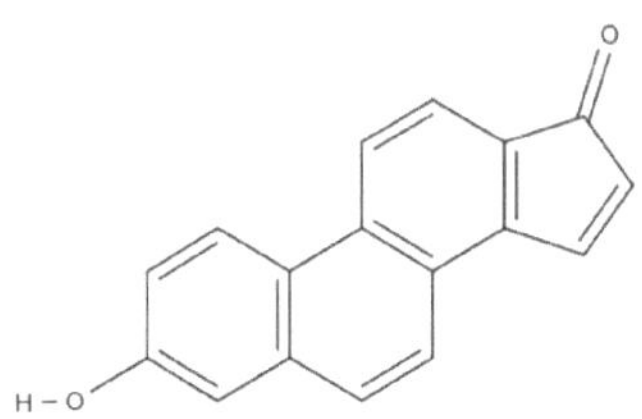

(c) Planar Steroid C-G Complementary

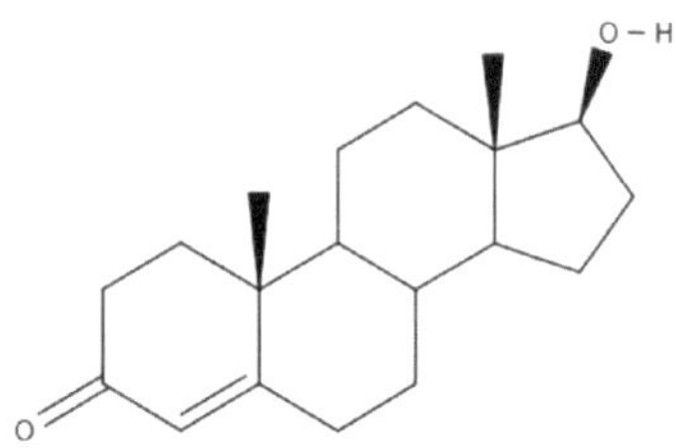

(d) modern-day Testosterone-Like

Figure 14.1: (a) The products of the origin of life function are the planar steroid molecule with cortisol-like shape and positioning of functional elements, and with significant double bond distribution; (b) the modern day version of cortisol-like steroid molecules has lost the double bond structure, and has just one double bond. Also added are methyl groups positioned around the hydroxyl group on the C-ring; (c) similar to cortisol-like, testosterone-like prebiotic, which is a product proposed from the unified complex, has double bond characteristics on all rings; (d) the modern-day version of testosterone has lost the double bond characteristics and has added methyl groups, similar to cortisol-like molecules.

In Figure 14.2, the side view of the comparative structures can be seen. There is a high degree of planarity for the steroids used in the origin of DNA and the genetic code, whereas modern-day steroid hormones do not have the same level of planarity. This will impact the ability of modern-day steroids to stack. There is also a loss of aromaticity. Moreover, the methyl groups added to modern-day steroid hormones significantly protect the hydroxyl group from interaction, which inhibits access of the hydroxyl group of the corticosteroids for direct binding strongly to dinucleotides.

Thus, there is evidence to indicate significant evolution of the molecular structure of the steroid molecules, so much such that in living cells, the steroids are unable to stack and to bind with high efficacy. Initially the planar steroids had the ability to

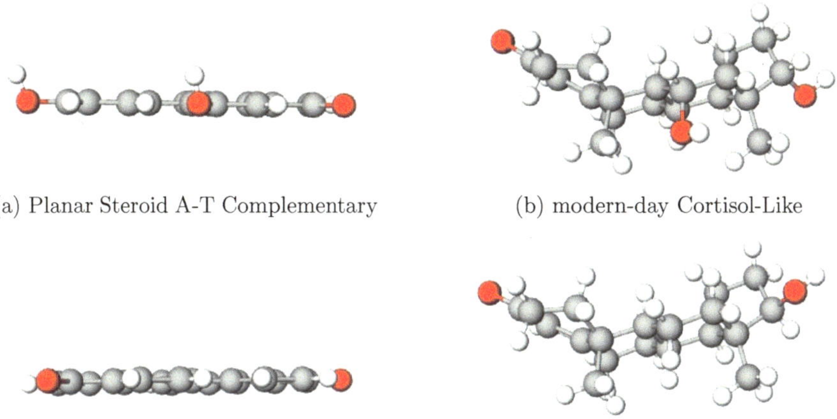

(a) Planar Steroid A-T Complementary

(b) modern-day Cortisol-Like

(c) Planar Steroid C-G Complementary

(d) modern-day Testosterone-Like

Figure 14.2: (a) Side view of cortisol-like prebiotic structure indicates planarity; (b) the side view of cortisol-like hormones indicates a loss of planarity due to the loss of double bonds and the gain of methyl groups; (c) Similar to cortisol-like molecules, testosterone-like molecules in prebiotic version, as synthesized from the unified complex and planar steroids, exhibit planarity; (d) loss of planarity and self-stacking capability is seen in the modern-day version of testosterone-like molecules in living cells.

stack and assemble to each other, which was necessary to form the DNA nucleotides, and ultimately to bind strongly with dinucleotides, which was necessary for transcription and replication. As these functions shifted to proteins, the planarity was not required and the strong binding characteristics with the dinucleotides was not required apparently. The evolution to a saturated C-ring also reduced the charge of the corresponding hydroxyl group in the A-T complementary planar steroid, thus reducing the strength of the hydrogen bond with a ketone group of the nucleobase thymine. Thus, there is a difference between the functionality of the original planar steroids and modern-day steroids, as indicated in Table 14.1.

This result is reinforced by evaluating other steroids used in living cells, as shown in Figure 14.3. The C-G complementary steroid hormones are all seen to have lost the double bond characteristic and the ability to stack and form strong bonds with the C-G pairing due to the methyl groups inducing steric hindrance. Interestingly, the estrogen series seems to have retained the aromatic nature on the A-ring, whereas the other C-G steroid hormones, such as testosterone and progesterone have lost the

Evolution of Molecular Structure	Result to Steroid Hormone in Living Cells
Loss of Aromaticity	Inability to self-stack due to loss of electron sharing capability; Loss of planarity; Loss of intercalation property
Gain of Methyl Groups	Reduced efficacy of direct binding capability to A-T and G-C nucleotides; Reduced efficacy of binding to thymine

Table 14.1: The evolution of molecular structure of the steroid hormones indicates that the first version had the same general shape and positioning of the functional groups. However, a loss of aromaticity and a gain of methyl groups resulted in a loss in the ability to self-stack and to exhibit strong direct binding to the A-T and C-G dinucleotides.

capability. It is also noted that the stronger anabolic steroids contain double bonds in the B and C rings. The A-T steroid hormones have also lost the double bond nature, while adding the methyl groups. One of the stronger synthetic compounds, Dexamethasone, has additional double bond characteristics in the A-ring, and a strong element to improve the binding to thymine.

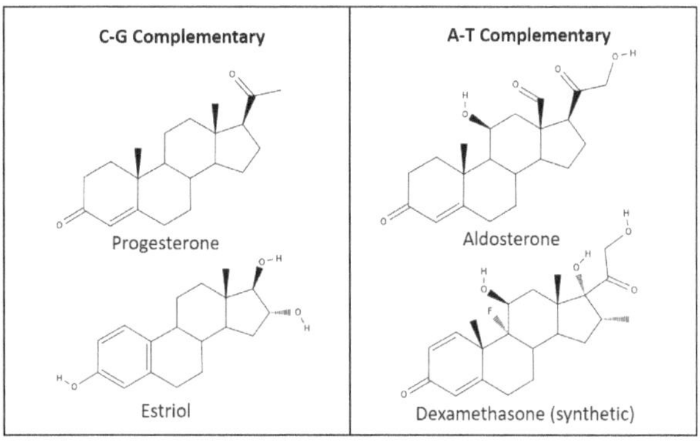

Figure 14.3: An analysis of other modern-day steroids all indicate a similarity in favor of an evolution in the molecular structure of the planar steroids that were first used in the derivation of DNA. Only estrogen related steroid hormones retained the aromatic ring and the hydroxyl group in the A-ring. Synthetic corticosteroids still retain the A-T complementary hydroxyl group, but do not have aromatic features.

It is noted that the interaction of steroid hormones with DNA nucleotides has shifted apparently to accommodate the transition of the phosphodiester backbone from PO_2H to PO_3^-. Since the A-ring contains a ketone group, it is unable to directly bond onto the backbone, with the exception being the estrogen series. Therefore, the interaction may be to the protein, in combination with direct binding of the C-ring through an intermolecular bond onto thymine, and the D-ring through hydrogen bonding onto the ketone group of the backbone. Thus, based upon the molecular models, a direct binding of steroid hormone to AT is conceivable. Since the complementary and symmetric nature of steroid hormones and DNA is still apparent, and even though the methyl groups create steric hindrance to reduce the efficacy of the potential steroid to AT hydrogen bond, there is still a path to bond the steroid molecule directly onto the DNA strands at the adenine-thymine coupling. Moreover, the relation would extend to testosterone-like steroids binding onto CG. It is necessary to distort the DNA double helix structure to pivot, but by means of transcription factors that seems feasible. Thus, there is a stable structure with hydrogen bonding of cortisol to thymine, as indicated in Appendix A.6 by the molecular models indicate that there is the potential for direct binding of cortisol to the A-T nucleotide. Therefore, it seems reasonable that there is not a complete loss of association between the steroid hormones and DNA, and transcription is not entirely mediated by allosteric effects through the glucocorticoid receptor. Consequently, modern-day transcription in living cells may involve the alteration of the DNA double helix through transcription factors, which position the A-T or C-G nucleotide pair for direct binding onto the complementary steroid hormone. The interaction appears to involve the protein integrated with the steroid hormone, which is also connected to the DNA nucleotides, generally in a dimeric pair. This results in improved stabilization at the promoter site, and improvement of the inducement of a nearby transcription bubble for association with RNA polymerase. The evidence for this mechanism of action is prednisolone versus prednisone. In my mechanism of action, the impact of a shift of a ketone to a hydroxy at the C-ring to enable activity is easily explained, modeled, and is predictable. Whereas with the allosteric theory on mechanism of action, the rationale to address prednisolone versus prednisone is not easily explainable. There is other evidence, as implied in the earlier discussions on the molecular primer, there is a strong predictive map of structure to function of the steroid hormones, both endogenous and synthetic, to the positioning of the bonds and elements.

14.2 Transitioning to tRNA, rRNA, mRNA from Planar Steroid Assemblies

The origin of life began from the planar steroid molecule, and the binding patterns associated with stacked assemblies. These design patterns, which could take place as in-phase and out-of-phase configurations, resulted in genetic material, an enclosed cell, and the molecular materials necessary for growth. Strategies for transcription, translation, and replication, as well as with genetic rearrangement were described, which all derive from basic properties of the self-assembly procedures associated with the planar steroid molecule. In this section, the evolutionary procedures are described which transition these rudimentary mechanisms of growth and reproduction to the modern-day protein-based mechanisms associated as the central dogma of molecular biology.

To produce tRNA, mRNA, rRNA, the mechanisms of stacking of planar steroid assemblies is coordinated with the genetic rearrangement procedures, particularly recombinant methods as described in Chapter 9.3, and the in-phase and out-of-phase binding phenomena as described in Chapter 12.1. The overall concept is that the genetic constructs arise out of binding configurations associated with the transcription process, which separates strands, thereby permitting strand-to-strand connections of planar assemblies originating from the ceiling associated with the interaction vessel. The cross-connections, and associated recombinant methods, result in genetic constructs that enable the specificity associated with tRNA, and the associated holding mechanism of rRNA, through in-phase and out-of-phase binding. The mRNA is produced via replication procedures as described in Chapter 8 and affiliated with the competitive driving force of transcription and replication.

14.2.1 tRNA Construct

To demonstrate the capability to form the genetic constructs of tRNA, rRNA, and mRNA from planar steroid assemblies, an example will be utilized. The example starts from separated strands of DNA as indicated in Figure 14.5. The distance between the strands can be larger, but here a rather short separation distance is utilized to keep the number of molecules in the demonstration more apparent to the example. As there is competition for replication and transcription, both processes are taking place at a respective strand. On one strand the transcription process is underway, as described in the previous chapters, in which interaction vessels are formed, as comprised of planar steroid molecules forming the walls and the ceilings. An opening is available every third nucleotide, which is closed by an amino acid, thereby com-

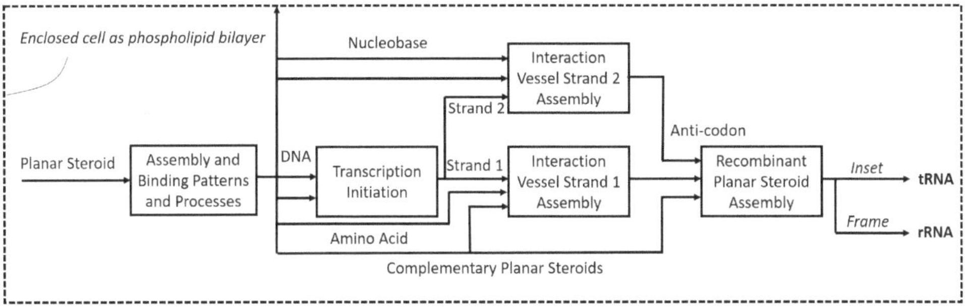

Figure 14.4: The process flow outline to start from an assembly of planar steroid molecules and produce the molecular constructs of tRNA, rRNA and mRNA, which comprise the central dogma of molecular biology to produce proteins from genetic material. The methods use the genetic fabrication procedures of replication and transcription, as well as recombinant methods, to produce the molecular structures. Of particular interest is tRNA, which has great specificity mapping amino acid to anti-codon: this match is achieved through a recombinant procedure linking the amino acid within an interaction vessel on one strand of DNA to its complementary match on the second strand undergoing preparation for replication.

pleting the chain, and defining the map from amino acid to trinucleotide codon. As we will see, the ceiling comprises a docking site for a projection towards the second strand, that will ultimately form the tRNA and rRNA constructs. On the other DNA strand, the replication process is underway, which was also described in the previous chapters. In this representation, complementary nucleobases in association with planar steroids occupy the available sites.

In Figure , the aligned dimers of the planar steroid molecule are indicated as projected off of the ceiling planar steroid molecule, which seeds the growth and alignment. These projections are parallel and project towards the second strand, ultimately coming adjacent to the nucleobases and planar steroids comprising the second strand. The distance traversed would normally be longer, but for discussion purposes, four dimers are indicated for each of the projections. The amino acids in this example are alanine and aspartic acid.

As discussed, the binding patterns of planar steroids result in a rich array of genetic and cellular constructs, including DNA and the cell enclosure, depending upon in-phase or out-of-phase binding patterns. The encoded tRNA molecule is another example of the binding pattern. Starting from the amino acid, a configuration is shown of which the phosphodiester bonds are developed which extends to the

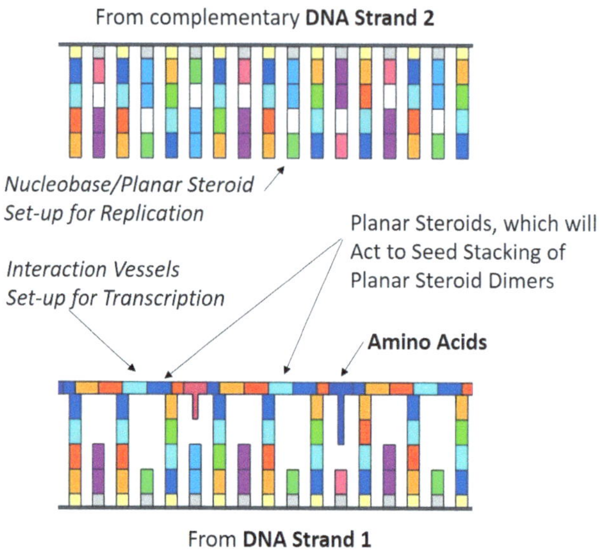

Figure 14.5: To describe the procedure to produce tRNA and rRNA from a planar steroid assembly, consider the separated strands of DNA induced by an strand opening initiation process. The situation in shown as a portion undergoing the transcription process, which involves the assembly of planar steroid molecules forming the walls and ceiling, and the amino acids centered over a nucleotide linked to its neighbors to form a codon. On the other strand, the replication process is favored, in which nucleobases are assembled with steroid molecules as described in previous chapters. The ceiling, which is comprised of an A-T complementary planar steroid, will act as a seeding for the docking and extension of aligned dimer groupings of planar steroids towards the second strand.

anti-codon and then back to the amino acid, with alternating covalent and hydrogen-bonding. This pattern will also result in nucleotides for the tRNA which through folding and secondary binding configurations will result in a genetic molecule.

To form DNA, a unified complex was achieved through a co-enzymatic process, and the same process is used in this example to form the DNA nucleotides at the molecules associated with the covalent bond, and the adjacent steroid molecule orientation defines the identification of the nucleotide, whether it is A, T, C or G, using the same encoding principles as those of DNA. There are some purines and pyrimidines, fifty percent on average, that are not in the same direction, and thus a

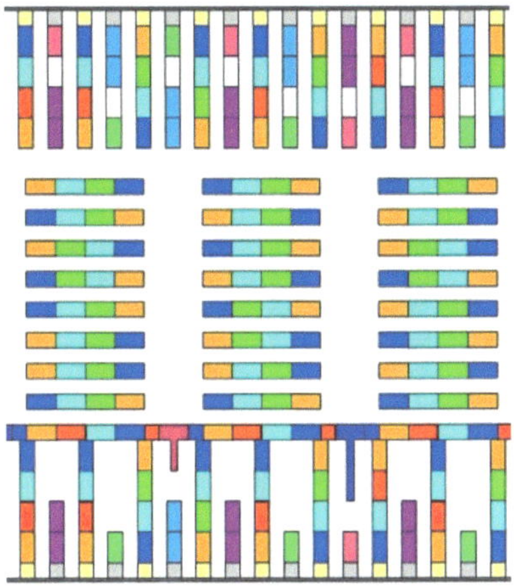

Figure 14.6: Dimer groupings of the planar steroid, are docked to the ceiling of the interaction vessel, and projected towards the second strand. There is a gap in between each projection of dimers, which transverses the separation between the two strands, and comes close in association with the second strand. Here an example is given of a truncated stack comprised of four dimers each.

driving force is created to release the steroid molecules, and thereby free the tRNA molecules, including the amino acid and the anti-codon. This will thereby complete the example of the processes, and the result is described next.

Through the tRNA formation process in this example, two tRNA molecules result as shown in Figure 14.9(a) for alanine and Figure 14.9(b) for aspartic acid. Both have an amino acid and an anti-codon, along with a coded strand. If this construction is feasible, then that would define the tRNA molecule, and with sufficient concentration be incorporated into the genome for replication. In case that the tRNA molecule candidate is not viable, such as in this example, it is not utilized and is degraded in the cellular environment.

It is noted that with this formulation of applying a planar steroid assembly to produce a construct that connects the amino acid to anti-codon, it is the case that the translation table after the integration of the construct with mRNA, which is the complementary phase to DNA, will therefore have the orientation of the modern-day. As indicated in Chapter 6, the translation table with the planar steroid directly producing amino acids off of the DNA strand had the opposite orientation to the modern-day translation tables, A instead of T, G instead of C. Thus the assem-

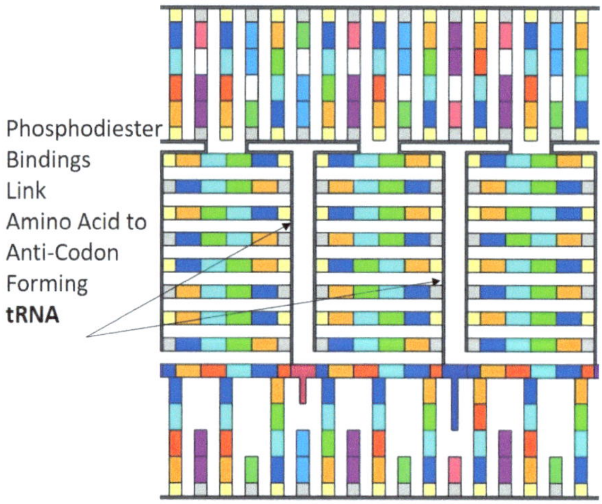

Figure 14.7: Through binding the phosphodiester chain along the ends of the dimer pairs of planar steroid molecules, patterns are feasible, which in this case results in tRNA (transfer RNA) extending from a binding of the amino acid from the interaction vessel of the first strand to its anti-codon on the second strand, and then back to the first strand, where it terminates. Thus, the total distance of this transversal process would be expected to be of the order to twice the nominal distance between separated strands of DNA.

bly provides the inversion to transition the planar steroid translation table to that associated with the central dogma of molecular biology.

14.2.2 rRNA Construct

Similar strategy as tRNA can be used to form rRNA with the main exception being that the amino acid is not present and the phosphodiester is bonded to surround two openings, such that the growing site and a staging position result for the elongation of the amino acids. The construction of rRNA is developed starting from the planar steroid assembly that projects to associate the two strands of separated DNA, as indicated in Figure 14.6. To form tRNA, the strand is connected into the amino acid, however in the formation of rRNA, the framework only is constructed. Continuing with the example and assembly used to demonstrate the formation of tRNA, this

Figure 14.8: As the planar steroid dimer pair is the same as that used in forming DNA through a co-enzymatic process, the same is the case for tRNA, where its coding will be along the extension of the dimers between the strands. Thus, as with DNA, a unified complex results will result, which consists of purines and pyrimidines aligned in their non-standard positioning, thereby creating a force to drive the planar steroids for separation, which will release the tRNA molecules including the amino acid and its anti-codon from the overall assembly.

process for rRNA is considered in Figure 14.10, which presents a trace induced by phosphodiester chain running through the configuration to connect two openings to the second strand of DNA.

After the trace is outlined as in Figure 14.10, the formative rRNA transitions to the unified complex, which undergoes the same co-enzymatic process of tRNA and DNA for planar steroid assemblies. This results in a unified complex comprised of nucleobases complementary and linked between the strands. There is an upper half, associated with strand 2, and a lower half associated with the connective planar steroid set, which defines the path for alignment of the tRNA strand as indicated in Figure 14.11. Thus, this construct induces a genetic profile to rRNA, which can result in its replication for duplication and optimization.

The extraction of rRNA from the unified complex results when the nucleotides are rotated into uniform position, resulting in a double helix, and ejecting the planar steroid molecules. The resultant trace is the connected rRNA strand shown in Figure

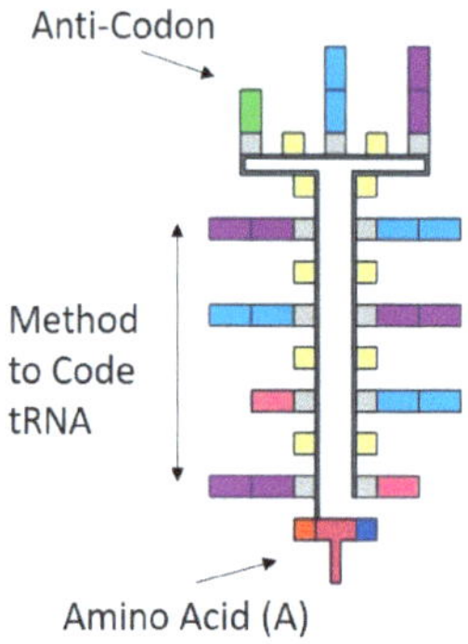

(a) tRNA template (A)

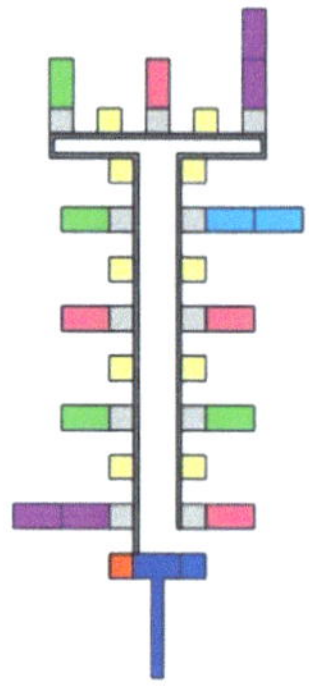

(b) tRNA template (D)

Figure 14.9: In this example, two tRNA candidates are shown. A nominal arrangement would have a longer chain and the coding would be different, but here the example should be clear to show that the planar steroid assemblies are capable of producing viable genetic constructs consistent with the central dogma of molecular biology. In (a), the alanine amino acid and its anti-codon are shown, and in (b), aspartic acid (D) its anti-codon are indicated.

14.12. The channels of rRNA in this example are capable of directing and aligning tRNA to the mRNA to match the anti-codon to the codon. Secondary structures are affiliated with rRNA to hold the mRNA in place while the transition occurs. Note that the rRNA contains the coding in this example that is the same for the tRNA along the connective area, which may aid in attracting and stabilizing the combined structure.

As is the case for tRNA, rRNA initially is constructed as rDNA and is thus incorporated into the genome. During the transcription process, rRNA and tRNA is the result, which is then put into practice. As will be seen, uracil replaces thymine, which enables removal of the construct when assisted through planar steroid molecules.

14.2.3 mRNA Construct

As for messenger RNA, its formation is covered in the chapter on replication, Chapter 8, and involves duplication during the transcription process with associated release due to the transition of the unified complex to a double helix. However, in this case, there is a pre-emptive disassociation of the mRNA strand from the transcript.

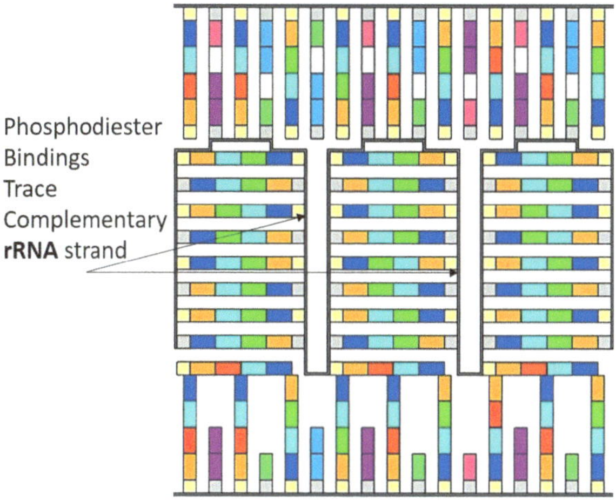

Figure 14.10: As a continued example of tRNA as shown in Figure 14.6, the framework to produce rRNA is demonstrated through a phosphodiester linkage of two adjacent channels which previously were used in the forming the tRNA, the framework for rRNA is traced within the planar steroid assembly.

This will be the case dependent upon a few factors, which includes the ability of the initiation site to maintain a transcription bubble, which will induce a defect at the initiation site, thereby enabling a single strand DNA construct to persist, while releasing the mRNA molecule. As will be seen in the subsequent section, the lack of a methyl group in uracil is of benefit in achieving separation during the transition from the unified complex to the separated strands. This result is represented in Figure 14.13.

14.2.4 Functional Comparison of Uracil and Thymine

In this subsection, the methyl group of thymine for DNA is compared for functional response to the hydrogen element of uracil for RNA, and it is shown that it is of crucial significance for the evolution from planar steroid assemblies to the central dogma of molecular biology. During the formation of DNA and its complementary planar steroid set, the unified complex, as an intermediate product, transitions to a double helix format by ejecting the intercalated steroid molecules. The methyl group on thymine participates in this process by steric hindrance promoting the removal of the

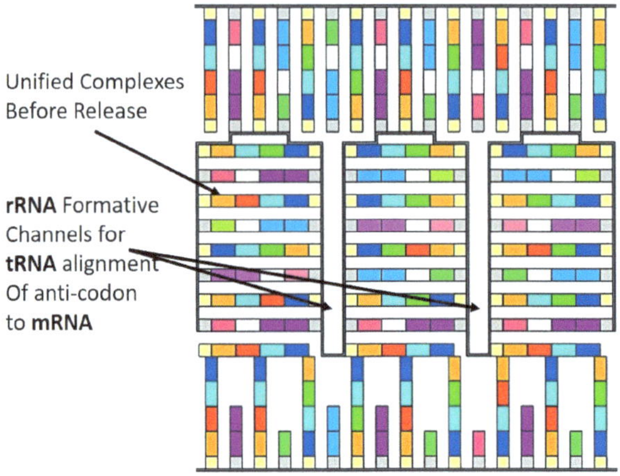

Figure 14.11: Similar to the case of tRNA and DNA, a unified complex results from the intermolecular activity of the planar steroid assembly set. This results in separation of the planar steroids into nucleobases, some of which are not oriented in the position compatible with the double helix. It thereby induces a genetic profile to rRNA, which can result in its replication when incorporated into a genomic construct.

adjacent planar steroid. However, in the case of the uracil, with the loss of the methyl group through the addition of an hydrogen element, the steric hindrance is removed, and uracil can rotate into position during the mRNA construction process without disturbing significantly the adjacent planar steroid. This suggests that mRNA is removable from the strand since the double helix is not formed, and the edge strands are decoupled due to the increase in hydroxy content of the sugar molecule along the strand.

In Figure 14.14(a) and (b), the intramolecular and intermolecular distances are presented in the comparison of thymine and uracil within a complex of planar steroid molecules. The rotation point is the nitrogen connecting the sugar to the nucleobase, and thus the distances are measured from that point. For thymine, the distance from the nitrogen to a hydrogen on the methyl group is 4.408 Å according ot the molecular simulator, and the intermolecular distance to the adjacent planar steroid is 4.169 Å. The nominal distance for stacked aromatic structure is approximately 3.8 Å, and this distance is roughly the same, although a little greater due to the slight tilt induced by the intermolecular binding to the hydroxyl group of the planar steroid. For the

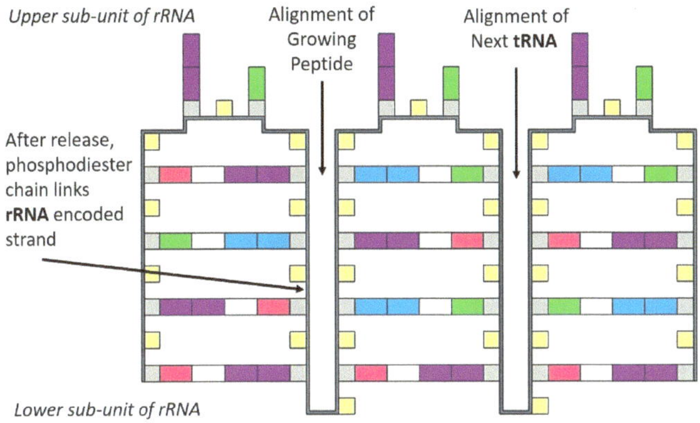

Figure 14.12: After the unified complex results in the rotation of the DNA nucleotides and the ejection of the intercalated planar steroid molecules, a strand of rRNA is available for processing the tRNA strand, directing and aligning it to the mRNA to match the anti-codon to the codon. Secondary structure on the upper portion, which does not have complementary nucleobase, is affiliated with rRNA to hold the mRNA in place while the transition occurs. Note that the rRNA contains the coding in this example for the tRNA along the stem area. This may help in stabilizing the molecules.

case of Uracil, which does not have a methyl group, the distance between the rotable nitrogen and the hydrogen, which is at the methyl position of thymine, is 3.373 Å, whereas the distance from the nitrogen to the adjacent planar steroid carbon is 4.212 Å. It is interesting the distances of thymine and uracil are equivalent amounts from the nominal distance between aromatic structures. Therefore, with rotation, thymine will sterically interfer with the planar steroid molecule, and drive it out of the complex to achieve a double helix. On the other hand, uracil during a rotation will not interfer and thus may simply be ejected from the complex, keeping the planar steroid molecule in place. Thus, with thymine, the double helix is formed with the complementary planar steroids ejected; whereas, with uracil, the complementary planar steroid molecules remain, while mRNA is ejected, thereby providing access for efficient translation processes.

The comparison of the functional significance of the differentiating structures between DNA and RNA is presented in Table 14.2. The replacement of the methyl group of thymine with a hydrogen for uracil is necessary for mRNA to be extracted from the unified complex, preventing the rotation into the double helix, which is

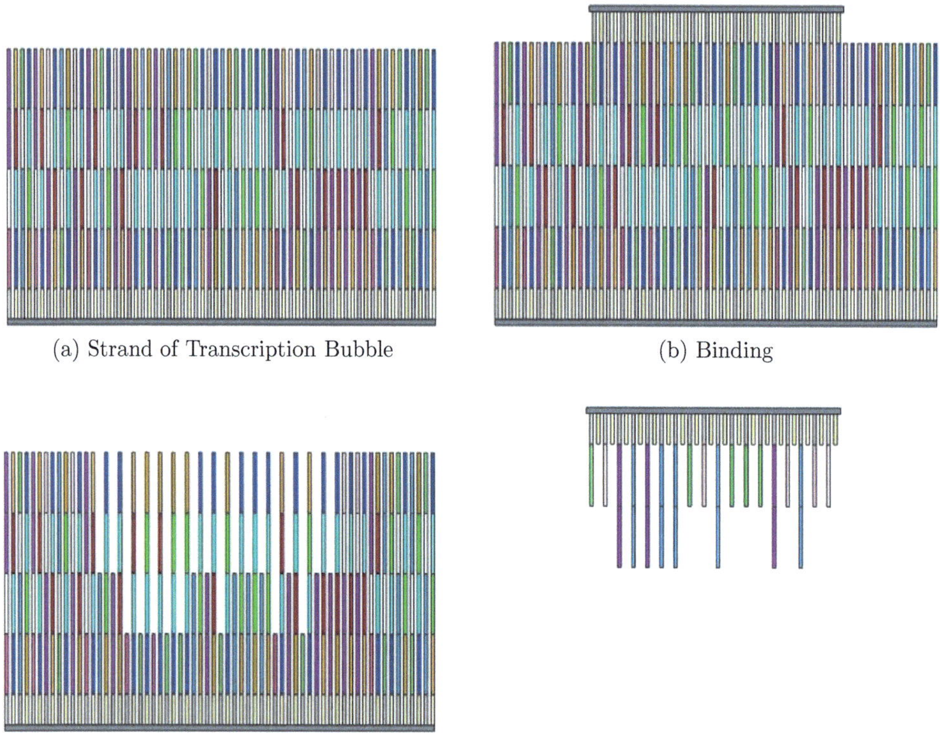

(a) Strand of Transcription Bubble

(b) Binding

(c) Remainder after Extraction

(d) mRNA Product

Figure 14.13: In this example, mRNA is formed. (a) Starting from a transcription induced single strand of DNA populated by ACGU as complementary, and configured through the planar steroid. (b) A select in-phase binding on the open strand, followed by separation (c) since uracil is used instead of thymine, resulting in (d) the mRNA product.

enabled by the methyl group of thymine. The addition of a hydroxy group to the sugar of RNA is helpful for removal of mRNA from the construct as it will increase solubility to the surrounding medium, thereby increasing the efficacy of the extraction process prior to the development of polymerase materials.

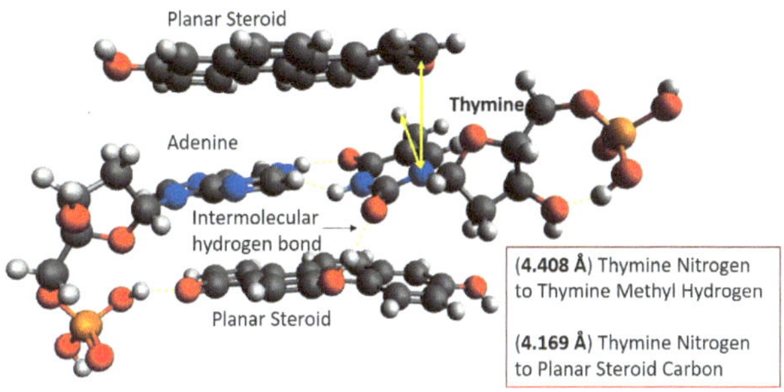

(a) Thymine to Planar Steroid Distances

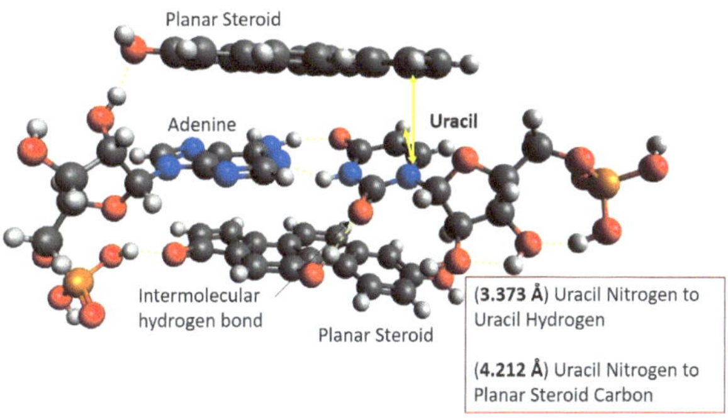

(b) Uracil to Planar Steroid Distances

Figure 14.14: The distances between the thymine and the planar steroid molecule is compared with that of uracil. (a) For thymine, the distances are calculed from the nitrogen connecting the sugar to the nucleobase with a hydrogen on the methyl group and a carbon on the adjacent planar steroid molecule. (b) For uracil, the distances are also calculated from the nitrogen connecting the sugar to the nucleobase, and the hydrogen at the corresponding position of the methyl group of thymine, and a carbon on the planar steroid molecule. Just a single phosphate group on either strand is shown for clarity.

14.2.5 Summary of RNA products

The planar steroid assembly and binding patterns enable a wide array of molecular constructs and life functions. An initial living cell is enabled, which through evolutionary processes transitions into the molecular structures known today such

Comparison	**Significance**
Replacement of methyl group on thymine with hydrogen	Loss of steric interaction between nucleobase and adjacent planar steroid during transition phases into the double helix preventing its formation when planar steroids are assembled
Replacement of hydrogen with hydroxy group on the sugar	Increased interaction of RNA with hydrophilic environment supporting its separation from the compiled strand of planar steroids and in-phase binding to RNA construction

Table 14.2: A comparison of the main differences between DNA and RNA, and the significance of these changes on the roles of DNA and RNA, and its interaction within the unified complex.

as steroid hormones and the transfer, ribosomal, and messenger ribonucleic acids, which are capable of producing proteins at a greater rate than the primitive construct. Further processing to develop chains of amino acids which further accelerate the processing rates, and strenghthen the cellular structure is the next step in the evolutionary procedure.

In Table 14.3, a summary is presented of the origin of the molecules that comprise the central dogma of molecular biology, all derived from planar steroid assemblies transitioning into the genetic information molecules to form proteins efficiently. The common link is the planar steroid assembly, which through a strand-to-strand recombinant framework results in the genetic sequence for tRNA, including its specificity and mapping of the anti-codon to the amino acid. In addition, the outline of the rRNA is associated with binding from the same strand-to-strand recombinant structure, thus inducing a frame for the structure that complements tRNA. The messenger RNA results from the competition between replication and transcription for access to the strands when open and available.

In sum, the planar steroid molecule and its assemblies and binding patterns established a rudimentary genetic framework for life origination, which had all the necessary functions for cellular life function including a cellular enclosure, genetic material, and functional operations of translation, transcription and reproduction. Through successive, progressive developments, a more efficient process and molecular construct was developed stemming from this basic function. In particular, RNA was developed for more efficient production of proteins, and proteins induced more efficient genetic flow and function, which both were adapted into the framework originated by the planar steroid assemblies and binding patterns.

Molecule		Origin
tRNA	Transfer RNA	Recombinant inset strand-to-strand binding of connective planar steroid assembly
mRNA	Messenger RNA	In-phase segment from replication in competition with transcription
rRNA	Ribosomal RNA	Recombinant framed strand-to-strand binding of connective planar steroid assembly

Table 14.3: After the origin of life is initiated through planar steroid assemblies, the competitive nature of transcription and replication initiated the development of molecular constructs that enabled efficiency through the use of different binding patterns of the planar steroid molecule sets to produce what is known as the central dogma of molecular biology, which comprises living cells. A key aspect is the development of a connective framework from the separated strands in which one strand prepares for transcription enabling the formation of the planar steroid assembly seeded by the planar steroid comprising the ceiling of the interaction vessel. The binding pattern forms tRNA and rRNA.

Chapter 15

Process to Elongate Peptides from Encoded DNA by Planar Steroid Assemblies

The capability of the planar steroid molecule to manipulate a polypeptide chain was given a preview in the discussion of the interaction vessel, which involved the formation of the peptide bond by a dimer carrier brought into position above the ceiling of the interaction vessel, thereby aligning the amine end of the peptide chain with the carboxyl of the amino acid retained within the interaction vessel. In this section, the details of that critical action, along with the basic unit interactions are indicated, which can then be used to form larger sets of interactions. While the genetic components of DNA are defined in the previous chapters, generated from the basis molecule of the planar steroid, there is still the need to manipulate the peptide product to form proteins, so as to induce the formation of proteins, which can then take over the manipulations in a more efficient manner. To do this, the planar steroid molecule is used both to hold and manipulate the protein material, as well as catalyze the peptide bond formation.

15.1 Interaction of Peptides with Planar Steroids

The AT-complementary planar steroid contains three functional groups: a hydroxyl group at the A and C rings, and a ketone group at the D ring. These functional groups can interact with amino acids through hydrogen bonding. In Figure 15.1(a), the single amino acid is stabilized between the C and D rings by hydrogen bonding with the amine group bonding onto the ketone group, and the carboxyl group of the amino acid binding onto the hydroxyl group at the C-ring of the AT-complementary planar steroids. For two amino acids, Figure 15.2(b) indicates that hydrogen bonding can also secure the di-peptide to the AT-complementary planar steroid through the C and D rings, using the same arrangement as the amine end hydrogen bonded to the ketone group, and the carboxyl group hydrogen bonded to the hydroxyl group. Since there are two amino acids, hydrogen bonding links the first and second amino acid.

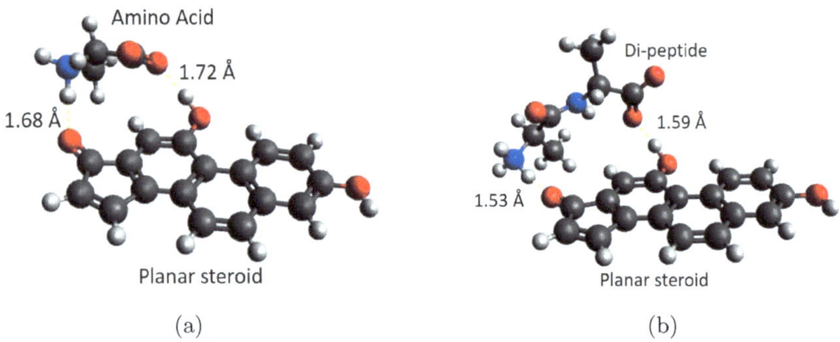

Figure 15.1: (a) Hydrogen bonding stabilizes the single amino acid two of the functional groups of the AT-complementary planar steroids. (b) Hydrogen bonding also secures the di-peptide to the AT-complementary planar steroid.

To extend the peptide across the planar steroid molecule, interacting with the end groups for the AT-complementary and for CG-complementary planar steroid molecules, comprises three amino acid units, thus two peptide bonds. Figure 15.2(a) illustrates this for three alanine residues. The protein takes the form with the residues projected outwards and alternating. For the three-peptide arrangement, it is not necessary for the hydroxyl group of the C-ring to form a strong hydrogen bond. The amine group bonds with the ketone group of the planar steroid, and the carboxyl group with the hydroxyl group at the six-carbon A-ring. The side view of Figure 15.2(b) indicates the interaction of the planar steroid and three-residue peptide.

To consider longer chains of peptides generated through a peptide elongation process, six residues and its interaction with the planar steroid molecule are indicated in Figure 15.3. The molecular simulation results show that the A-T complementary planar steroid deploys three hydrogen bonds to the peptide for each of its three functional groups to position the growing peptide. Ultimately, this configuration is used to association with the amino acid contained within the interaction vessel, and thus the angle of the charged amine end-group is of interest, as well as the nearby carboxyl group. Driven by the planar steroid molecule, this arrangement provides a dual function of alignment with the amino acid to be added and providing an oxygen

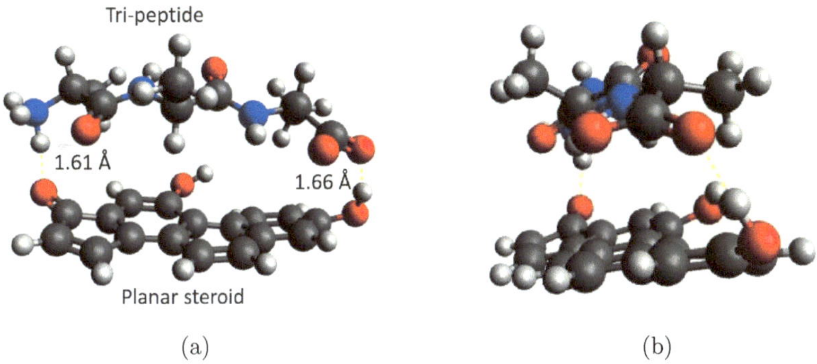

Figure 15.2: (a) The minimum number, three, of peptide bonds needed to traverse a planar steroid molecule with hydrogen bonding at either end is indicated. Note that the amine group bonds with the ketone group of the planar steroid, and the carboxyl group with the hydroxyl group. The methyl group is used as the amino acid residual, which corresponds to alanine. (b) Side view shows the interaction and hydrogen bonding at ends.

to participate in the formation of the water molecule associated with the peptide bond. The form of the polypeptide as it exits the planar steroid sweeps away from the arrangements and thus will not interfere with the elongation process.

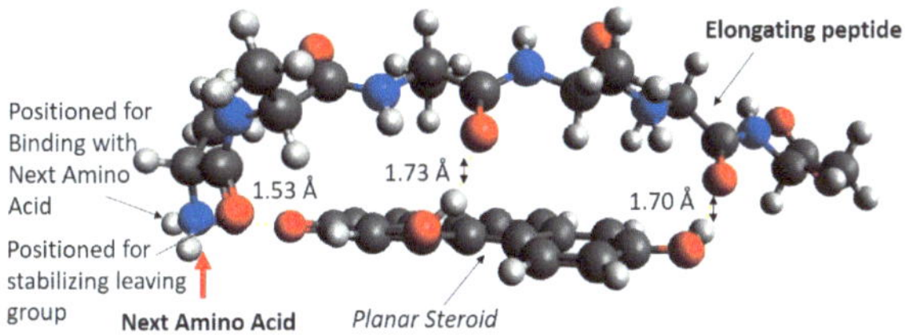

Figure 15.3: The elongated peptide is also stabilized to the AT-complementary planar steroid which uses three hydrogen bonds to position the structure. Note the angle of the amine end-group which and the nearby carboxyl group. This will enable alignment and the formation of the peptide bond. Also note that the peptide is swept away from the elongating structure after it exits the AT-complementary planar steroid.

The strategy to form peptide bonds through the use of planar steroid molecules is indicated in Figure 15.4, in which two of the three hydrogens associated with the charged amine group are associated with hydrogen bonding to the functional ketone end-group of the five-carbon ring of the planar steroid molecule, which will stabilize and align the amine group. Thus, the third hydrogen atom is available and aligned with the elongating chain, which will attract the carboxyl group for attachment of the next amino acid. Moreover, the hydroxyl group of the carboxyl group of the next amino acid will associate with the associated and align with the adjacent ketone of the first peptide. This combination of alignment mechanisms will enable the formation of the peptide bond and the associated water molecule product.

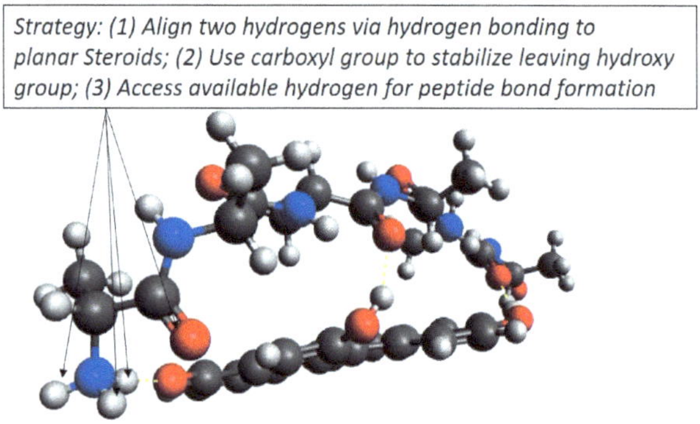

Figure 15.4: The basic strategy to form the peptide bond: two of the three hydrogen atoms of the charged amine group are associated with hydrogen bonding to the planar steroids. The third hydrogen atom is available and aligned and will attract the carboxyl group of the next amino acid. The hydroxyl group of the carboxy group will associate with adjacent ketone of the first peptide. The formation of the peptide bond and the associated water molecule products are the result.

15.2 Elongation Process by Planar Steroid Interaction Vessels

A feed stream for amino acids via the interaction vessel assembly mechanism is associated with the methodology of forming peptide bonds through alignment of the planar steroid molecule, and stabilization provided from the surface of the planar steroid molecule. To demonstrate the approach, the simulation model of the interaction vessel is depicted in Figure 15.5 for a TGATGA sequence, which is associated with the amino acid threonine. It shows that the amino acid binds onto both the planar steroid molecule at the ceiling, and the planar steroid walls. Threonine is in position to directly interact with the second nucleotide comprising the codon triplet, which is guanine. The spacing of the interaction vessel is appropriate as the length of the planar steroid molecule is equal to twice the spacing from one wall to the next, and thus can cover two vessel openings and thereby retain only one opening for the insertion of the amino acid, as was described in previous sections. Also, as a reminder, the length of the amino acid from carboxyl group to amine group is approximately equal to the opening region of one vessel, and thus the combination with the planar steroid will establish the code of three nucleotide to one amino acid observed for the translation map of nucleotide sequence to amino acid. Thus, since the translation map is clear, the mechanism to form peptide bonds in an elongated chain is needed.

Continuing with the discussion of the example for peptide formation, the molecular simulation model of Figure 15.5 for the capture of two amino acids in the TGATGA nucleotide sequence, which corresponds to the amino acid sequence TT. This is representative of a transcription bubble which exposes the strand, and after assembly of the planar steroid molecules and amino acids, it is ready to begin peptide elongation by moving the amino acids from the upstream segment to form a peptide bond with the downstream amino acids, all contained within the interaction vessels. There are two codons in this example, and the association of the amino acid to the planar steroid is through hydrogen bonding at one end to the carboxy end of the amino acid, and at the other end by hydrogen bonding of the amine group of the amino acid to the ketone group of the A-T planar steroid.

The initiation of the peptide bond formation process is indicated in the simulation result of Figure 15.6. The integration of a second planar steroid aligned with the ceiling of the interaction vessel is the key step. Here it is shown with the elongating peptide associated through hydrogen bonding to the three functional groups of the AT-complementary planar steroid. This alignment process of forming a dimer configuration with the ceiling aligns the amino acid in the interaction vessel with the amino acid, which was formed from the adjacent upstream interaction vessel site. The planar steroid of the ceiling associates through hydrogen bonding with the second of the three hydrogen atoms associated with the charged amine group. The charged nature of the amino acid associated with the elongating peptide chain will attract the carboxyl end of the amino acid in the interaction vessel which will rotate towards the elongating chain, which is in perfect alignment.

As mentioned, the binding of the ceiling onto the available amine group, which represents the most recent amino acid whose source is the adjacent upstream interaction vessel, initiates the peptide bond formation process. The dimensions associated with the binding process of the ceiling to the amino acid is indicated in the simulation result of Figure 15.7. This shows that the planar steroid ceiling undergoes

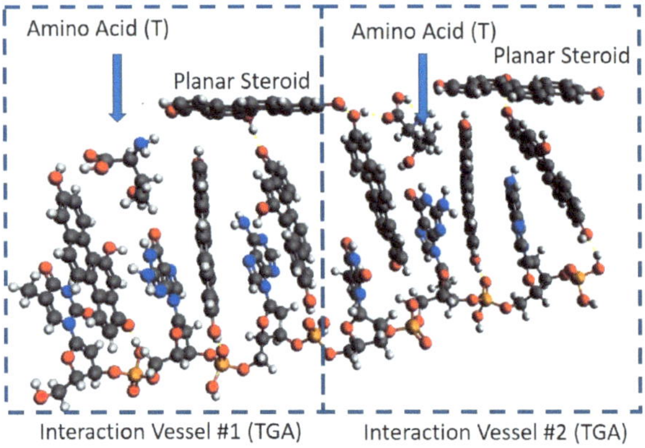

Figure 15.5: A portion of the transcription bubble is depicted, with two amino acids assembled within the interaction vessel, prior to peptide elongation. The interaction vessel is formed by planar steroid molecules comprising the walls, and an A-T complementary planar steroid molecule comprising the ceiling. In this example, two interaction vessels are shown for TGA sequence and the threonine (T) amino acid is indicated, which corresponds to this codon.

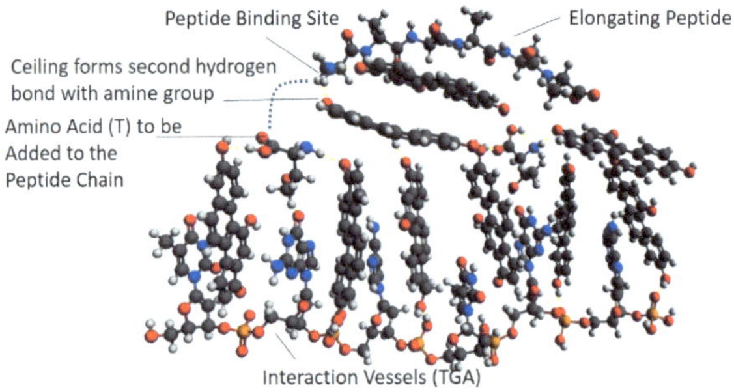

Figure 15.6: The procedure for peptide elongation is indicated. A dimer is formed on the ceiling with an aligned complementary planar steroid associated with a polypeptide. The amino acid in the first interaction vessel is attracted to the charged amine group, and the planar steroid ceiling is associated with the dimer group thereby releasing the amino acid for transition to the charged amine group.

hydrogen bonding to the charged amine group hydrogen atom, and thus enables a single hydrogen atom for bonding at the designated angle, consistent with a peptide bond formation. Thus, there is only one hydrogen atom of the amine group available for bonding, and it is in alignment consistent with a peptide chain.

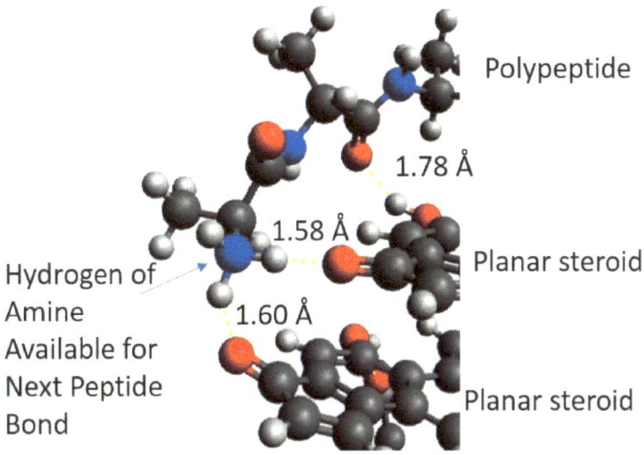

Figure 15.7: The planar steroid ceiling undergoes hydrogen bonding to the charged amine group hydrogen atom, and thus enables a single hydrogen atom for bonding at the designated angle, consistent with a peptide bond.

The amino acid of the interaction vessel is attracted to the available hydrogen atom of the amine group, transported to the site in association with the displacement of the ceiling of the interaction as it comes into association with the elongating peptide. The overall figure indicating the result of the transportation of the amino acid within the interaction vessel to the binding site of the elongating peptide chain is indicated in the simulation result of Figure 15.8. The planar steroid molecules have hydrogen bonding to two of the hydrogen atoms of the amine group, and the amino acid has hydrogen bonding via the carboxyl group to the third hydrogen atom. Moreover, the hydroxy group of the carboxyl group of the amino acid is hydrogen bonded onto the ketone group of the first peptide, which is the most recent addition to the elongating peptide chain.

The dimensions of the association of the amino acid with the peptide binding site are shown in Figure 15.9 for the molecular simulation. The result indicates that the distance of the hydrogen bond between the planar steroid ceiling ketone group and a hydrogen element of the amine is 1.78 Å, the distance of the hydrogen bond between the planar steroid carrier molecule ketone group and a hydrogen atom of the amine is 1.66 Å, and the distance from the ketone group of the amino acid to the hydrogen atom of the amine group is 1.49 Å. Thus, all are in tight association at the peptide coupling site.

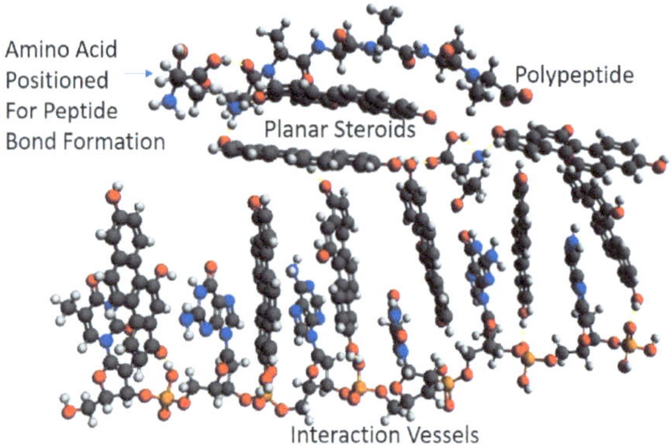

Figure 15.8: The amino acid hydrogen bonded onto the amine group of the polypeptide. The planar steroids hydrogen bond two of the hydrogen atoms of the amine group, and the amino acid is hydrogen bonded from the carboxyl group onto the third hydrogen atom. The hydroxy group is hydrogen bonded onto the ketone group of the first peptide.

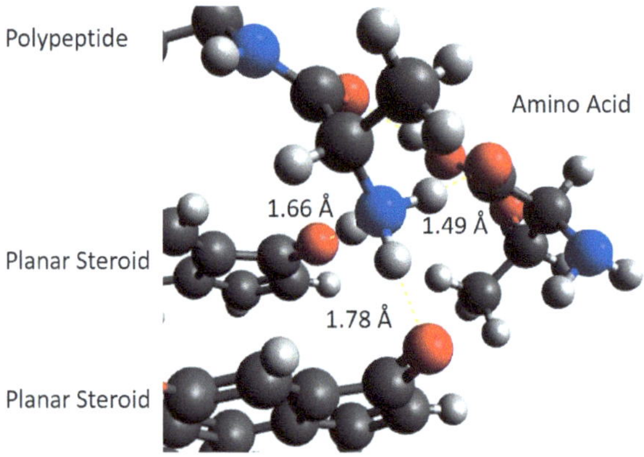

Figure 15.9: Hydrogen bonding associates the amino acid to the amine group, stabilized by the two planar steroid molecules to enable the peptide bond to form.

It is also important to evaluate the dimensions associated with the leaving groups associated with the water product of the peptide bond formation process. As shown in Figure 15.10, the leaving group associated with the hydroxyl group of the amino acid is in hydrogen bonding with the ketone group of the amino acid that is to be bonded by the amino acid peptide formation process. The hydrogen bond distance is 1.81 Å, and is thus aligned, stabilizing the amino acid molecule for interaction.

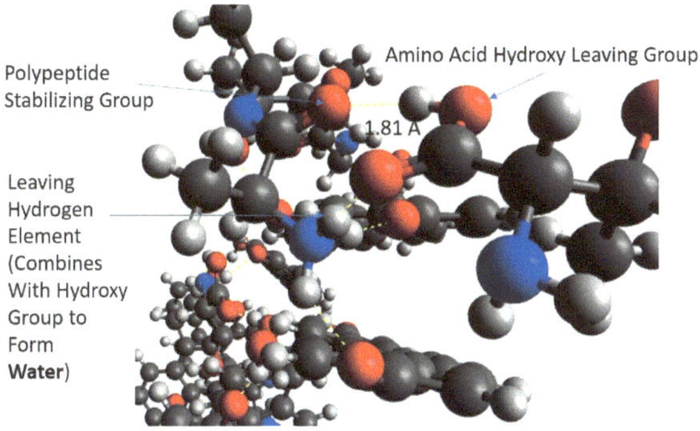

Figure 15.10: The leaving groups associated with the amine group and the hydroxy group of the amino acid are in close proximity and well positioned for the peptide bond to take place.

Moreover, as indicated in Figure 15.11, there is proximity of the leaving element of the hydrogen atom of the amine to the hydroxyl group of the reactant amino acid to be added to the elongating peptide chain. This enables bond formation and water product formation, which will not disturb the planar steroid assembly as it is hydrophobic on the interior. The angle of the bond is determined through this configuration as well.

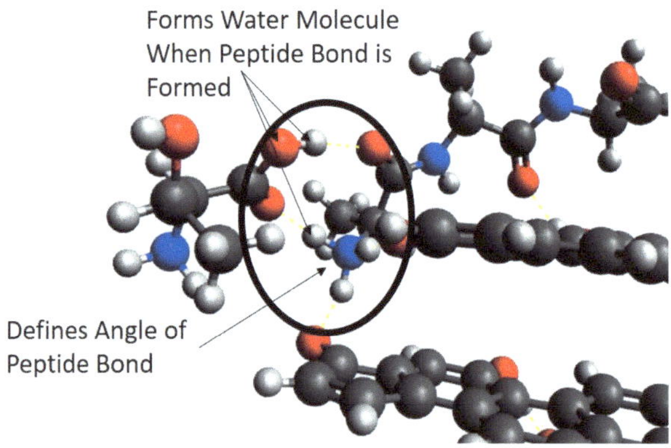

Figure 15.11: The angle of the next peptide element is defined by the position of the hydrogen atom, and therefore defined by the positioning of the two planar steroid molecules which isolate the available hydrogen atom of the charged amine group. The water molecule is formed from the close proximity of the elements during the bonding process.

Chapter 16

Encoding DNA from the Protein Sequence

Even with the advancements described in the preceding chapters on the generation of the genetic code from planar steroid assemblies and binding patterns, there is still the nagging question of where all of the information came to encode all of the protein structures associated with the cell. These include both structural features, such as the cytoskeleton, as well as chemical features, such as enzymes. It seems far too inefficient that all sequences of nucleotides that encode for efficacy happened all by random chance. In this chapter, I lay the foundation to solve this problematic issue and then in the following chapter through molecular simulation models demonstrate its implementation by several examples.

While the current central dogma of molecular biology describes the transformation of information from nucleic acid compounds into protein structures, it is not unreasonable to consider that to generate the information in the first place, that the current central dogma of molecular biology was not the case: that is, to initiate the process effectively, there was information flow from protein to DNA, and in this section, I show how to do it. Previously, I described the genetic code as derived from the interaction vessel, which associates a strand of DNA into sequences of three nucleobases, each in association with its complementary planar steroid. The amino acids reside in the interaction vessel separated by a ceiling, which primed the peptide bond formation. In this chapter, I demonstrate the interaction vessel also work in reverse, meaning that the residue of the amino acid can trigger the formation of its encoded DNA. This procedure can be considered an inverse transcription process. Thus, the central dogma of molecular biology, while true today, was not always the case. And, it will be seen, a significant volume of information to support and operate the cell was generated through this inverse transcription process.

16.1 Planar Steroid Mechanism to Synthesize Proteins while Encoding DNA: Inverting the Current Central Dogma of Molecular Biology

To explain the process of elongating a peptide while encoding the elongation into DNA as it occurs, an example is utilized of a polypeptide with threonine residue at the second peptide position with all other residues alanine. In association with the AT-complementary planar steroid, which exhibits hydrogen bonding at three sites, the nitrogen group is positioned at the five-carbon ring position for formation of a peptide bond, as described in the preceding chapter. On the other end at the six-carbon end ring, the A-ring, the polypeptide chain is free to continue growing during the elongation process. To separate peptide elongation with encoded DNA elongation, the strategy is indicated in the model of Figure 16.1, which presents that at the first residue of the peptide chain is engaged in the formation of the peptide bond in the peptide elongation process, and at the second residue, there is engagement in an encoding process to elongate DNA strand. The strategy is that the residue at the second site is read by the interaction vessels forming a codon and written into a DNA strand by formation of the phosphodiester-sugar bond. Both the peptide chain and the DNA strand are thus elongating at the same time in synchronicity to each other with the DNA strand elongating three nucleotides for each residue added to the elongating peptide. Thus, after the formation of the peptide bond at the first site and the tri-nucleotide encoding based on the residue at the second site, the units increment and a new peptide bond is formed and the residue of the former first site is incremented to be read by a new tri-nucleotide codon.

The feasibility of associating interaction vessels with the protein residue to reverse encode DNA is shown in Figure 16.2. This figure shows the result of a simulation of interaction vessels associated with the second residue of the polypeptide chain, which is Threonine and the codon is TGA, which was studied in the previous chapter, and whose interaction vessel construction was described in the novel translation table of Chapter 6. The procedure involves forming the interaction vessels in the tri-nucleotide sequence, along with the corresponding planar steroids, that can form a stable structure with the second residue, and as will be seen. After the phosphodiester backbone is formed, the encoding procedure is completed for the second residue. In so doing, the planar steroid of the third wall connects with the hydrogen atom of the nitrogen group associated with the amine group of the first residue. Consequently, the next amino acid can bind onto the amine group of the first residue and the process

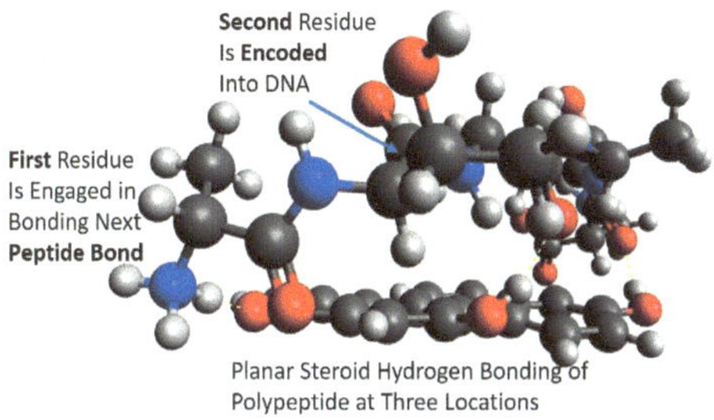

Figure 16.1: The first residue of the peptide chain is engaged in the formation of the peptide in the elongation process. The second residue is engaged in an encoding process, in which the residue is read by the interaction vessels forming a codon and written into a DNA strand, and thus is affiliated with a DNA elongation process. Both the peptide chain and the DNA strand are elongating at the same time in coordination to each other with the DNA strand elongating three nucleotides for each residue added to the elongating peptide.

can continue in synchronicity. This process is then repeated as the polypeptide chain elongation process continues propagating the chain about the AT-complementary planar steroid.

As shown in Figure 16.3, extensive hydrogen bonding keeps the entire unit together, as well as enable its functionality for encoding and for peptide bonding. The hydrogen bonding of the two planar steroid molecules are 1.83 Å corresponding to the third wall of the interaction vessel and 1.62 Å for the AT-complementary carrier planar steroid, which has 1.72 Å hydrogen bond distance for the functional group at its C-ring. The distance of the second wall to the nitrogen group of the second peptide is 1.91 Å and the carboxyl group of the first planar steroid is 1.81 Å to the ketone group of the second peptide residue. The nucleobase guanine is held by hydrogen bonding to threonine at 1.91 Å. There is one available hydrogen atom for bonding to form the next peptide bond. Note that the third wall, that is the third planar steroid of the codon encoding for the second residue, has a similar function of the ceiling of the interaction vessel when encoding takes place in the forward direction, from the DNA strand to the amino acid. Both planar steroids are needed to isolate the single hydrogen atom, aligning it with the next amino acid.

Another view and labeling of the purposes of the complex is considered in Figure 16.4. The first peptide group is available for interaction with the next nucleotide, and the third wall of the codon, which one will recall for the forward case of the interaction vessel was for alignment of the ceiling, has the role of triggering the peptide

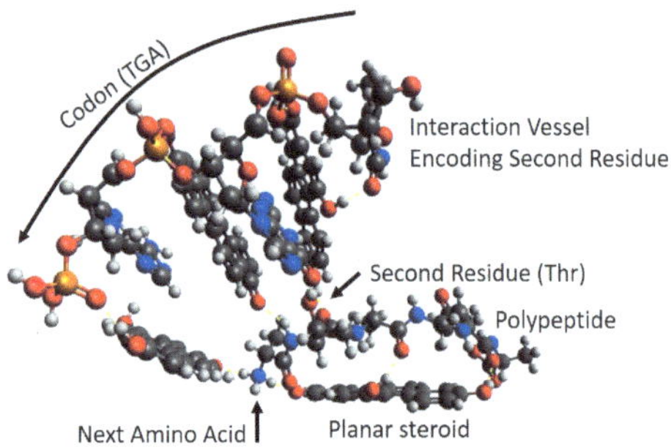

Figure 16.2: The interaction vessels are associated with the second residue of the polypeptide chain. The procedure involves forming the interaction vessels in the tri-nucleotide sequence, along with the corresponding planar steroids, that can stabilize the residue. The phosphodiester backbone is then formed, thus the encoding procedure is completed for the second residue, and in so doing, the planar steroid of the third wall connects with the hydrogen atom of the nitrogen group associated with the amine group of the first residue. Consequently, the next amino acid can bind onto the amine group of the first residue and the process can continue. The example here shows the TGA tri-nucleotide codon for the threonine residue, corresponding to the translation table associated with the interaction vessel in Chapter 6.

bond formation process, while also necessary for the formation of the phosphodiester linkage, thereby providing timing sequence for the process. The second residue, including its adjacent oxygen and nitrogen elements, provide the coupling points for the second and first planar steroid walls. The ketone group associated with the first residue is also engaged in the process to align the incoming amino acid. Thus, the functional groups for both the elongating peptide and the interaction vessel encoding for DNA are used.

The organization of the Peptide-DNA Synthesizer-Encoder system is apparent in the van der Waals space filling molecular simulation model of Figure 16.5. It shows, for example, that the hydrogen atom and the ketone group are accessible for interaction with the next amino acid to form a peptide bond. The planar steroid molecules, which are connected to the charged amine group of the attractive site, are angled to project the reactive site towards the amino acid. The ketone group is available or alignment of the incoming carboxyl group from the amino acid, which will assist with water formation during peptide synthesis.

The alignment and binding of the next amino acid onto the first residue of the

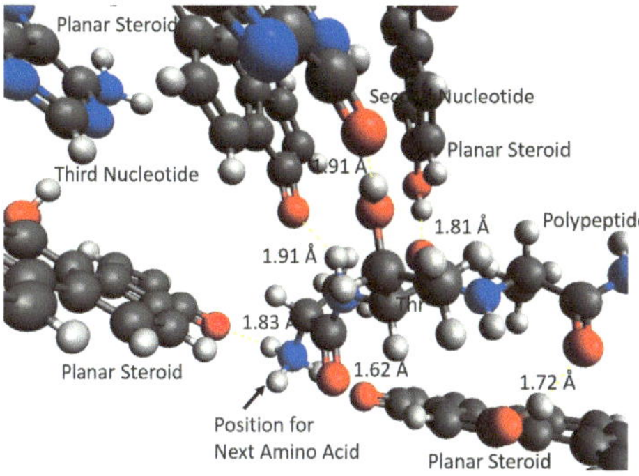

Figure 16.3: Extensive hydrogen bonding keeps the entire unit together. Two of the hydrogen atoms of the amine group associated with the first residue are in binding with the planar steroid oxgen atoms at the five-carbon position. Additional hydrogen bonding is shown for the two walls of the interaction vessel on the amine and oxygen group of the second residue. Other hydrogen bonding is associated with the nucleobases binding toy the planar steroids or the second residue.

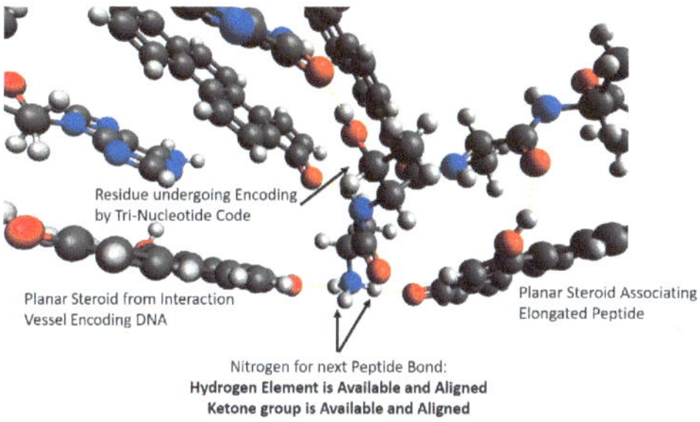

Figure 16.4: The first peptide group is positioned for interaction with the next amino acid to be added to the elongating peptide. Only one hydrogen atom is available for interaction and it is aligned perfectly. Also available and aligned is the ketone group of the first residue. The planar steroids bond onto the other two hydrogen atoms of the charged amine group.

elongated peptide chain is shown by the molecular simulation of Figure 16.6. There is a clear path to form the peptide bond. Thus, the overall system results in a coordinated mechanism for writing, reading and encoding information in which peptide elongation occurs at the same time as the DNA elongation. Upon the formation

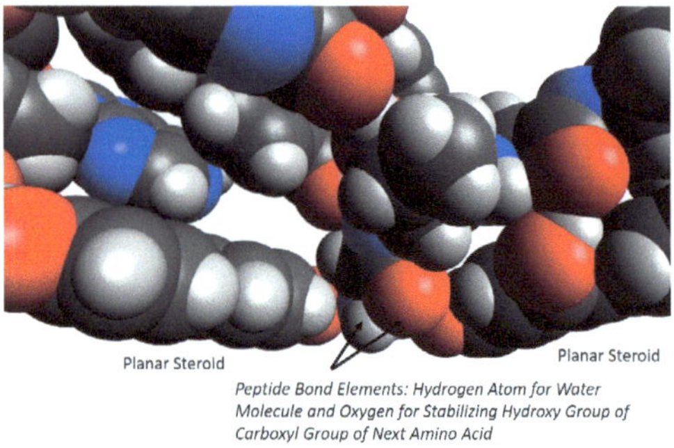

Figure 16.5: *The van der Waals space filling model indicates that the hydrogen atom and the ketone group are accessible for interaction for the formation of a peptide bond with the next amino acid.*

of the peptide bond, the next codon is formed including its full interaction vessel, and the process repeats. Not shown in the picture is the formation of the full DNA strand, which occurs downstream in conjunction with the planar steroid molecules. The intercalated planar steroids travel with the phosphodiester-sugar linkage to form the full DNA strand using the techniques described in Chapter 8.

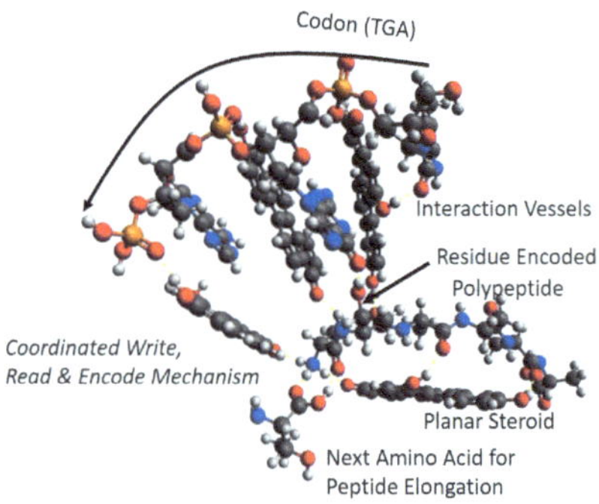

Figure 16.6: *The binding of the next amino acid onto the first residue to form a peptide bond. Consequently, it is a coordinated mechanism to write information as peptide chains while at the same time encoding the writing into a tri-nucleotide codon corresponding to the amino acid residue. Upon the next amino acid forming peptide bond, the next codon is synthesized in tandem.*

The dimensions associated with the complex after the amino acid presents to the binding site is indicated in Figure 16.7. The hydrogen bond distances associated with the three hydrogen elements are 1.62 Å to the presenting amino acid, 1.62 Å to the carrier planar steroid and 1.83 Å to the third wall of the interaction vessel, encoding for the DNA strand. The distance between the hydroxy group of the amino acid and the ketone of the first residue is 1.63 Å. The appearance of the amino acid for binding onto the chain thus does not disturb the overall complex. Also noted is the entrance and exit sites have clearance capability to all unobstructed access and production.

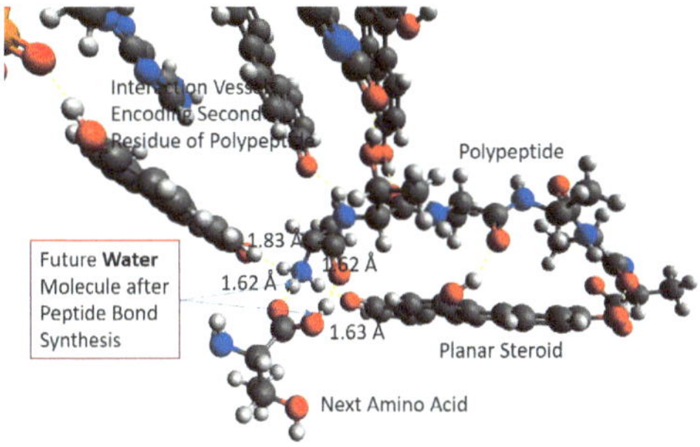

Figure 16.7: Hydrogen bonding readies the structure for peptide bonding with the amino acid binding onto the first amine and carboxyl segments of the first peptide group. Planar steroids bind onto the other two hydrogen atoms of the amine group of the first peptide. The future water molecule components are in close association.

As this procedure involves an inverse transcription process, it is interesting to consider the bidirectionality of the genetic code through the interaction vessels. The construction of the codon mapping was done by associated stable chemical and structural arrangements of nucleobases and associated planar steroid molecules, but the developments considered the nucleobase first and examined the amino acid as associating with the nucleobase. Here the reverse is examined, and considers the amino acid residue, and inquires about the nucleobase triplet, and complementary planar steroid molecules, that will result in the most stable structure. The structural and chemical characteristics of Tables 6.1 and Table 6.2, associating interaction vessel characteristics and amino acid candidates would be bi-directionality, with the notable exception of glycine. However, for this case, without a driving force residue, the wall would be comprised of CG-complementary planar steroids, which do not have mid-molecule functional group, and the shorter single-base nucleobase would be selected to improve dimer alignment of the planar steroid walls. Thus, although further study is needed on the specificity of the interaction vessel to amino acid relation, it is feasible to consider that the bidirectionality of the genetic code is maintained.

16.2 Binding and Encoding Formative Peptide Constructs

In the next chapter, the method to encode into DNA a wide range of peptide structures will be shown, but here to finish the introduction to inverse transcription, I wanted to indicate that the Peptide-DNA Synthesizer-Encoder mechanism should reside or rest on a stable structure, and have a source for the amino acid chain that can accumulate a chain of amino acids, and is stable such that the assembly and transfer process can occur. One such structure is indicated in Figure 16.8, which provides a stable site to secure the coordinated synthesis and encoding mechanism while accumulating and feeding amino acids in a structured format. Consequently, there is the situation in which the amino acid sequence which conforms to the docking site will be encoded, and thus for this case, one would expect the free flowing amino acids that assembly along the planar steroid will produce a polypeptide whose residuals will not conflict with the structure from which the sequence was derived. Therefore, the encoded DNA sequence would be capable of producing replicated structures from the one base structure.

As was described for the amino acid presentation, the docking of the peptide-DNA synthesizer-encoder assembly does not restrict the entrance and exit point for the elongating peptide chain, as indicated in Figure 16.9. The system is appropriately designed such that the amino acid entry site has clear access and the exit point for the elongated chain has clear exit point and is directed in a direction similar to the entry point. The entrance chain into the peptide-bond synthesis sits is not impacted, and the DNA strand elongation process is far removed from the peptide bond synthesis site.

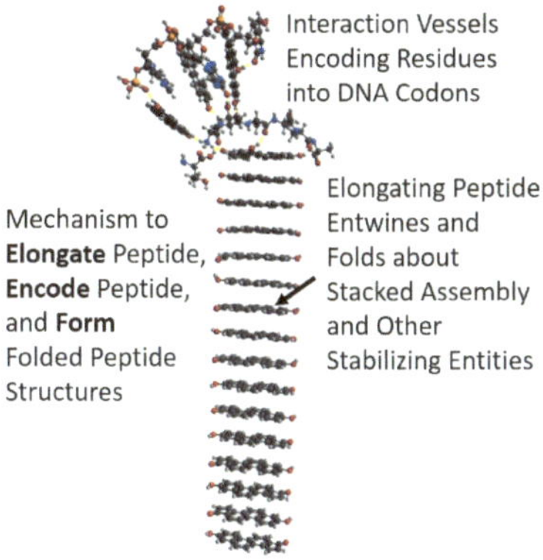

Figure 16.8: The source of the amino acids for the coordinated synthesis and encoding mechanism are structures which are capable of accumulating amino acids in a structured format, and provide a stable docking site for the lead planar steroid, or dimer pair. Consequently, the amino acid sequence will conform to the set of amino acids associated with the areas adjacent to the synthesis site. The proteins that exit the synthesis site entwine and bond onto the structure from where the sourced amino acids started.

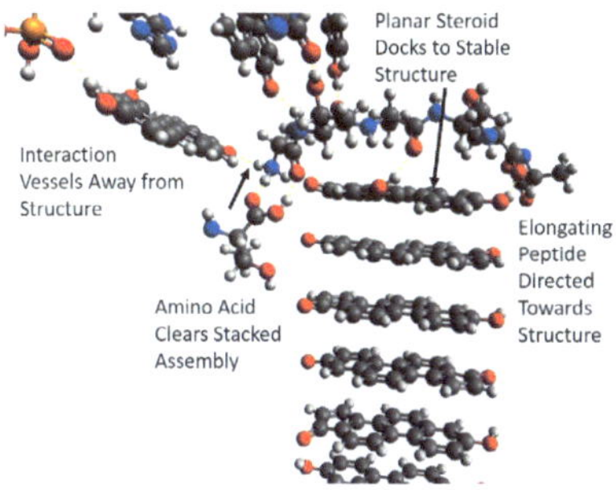

Figure 16.9: The system is designed appropriately such that the amino acid entry site has clear access and the exit point for the elongated chain has clear exit point and is directed in a direction similar to the entry point. The interaction vessels are away from the synthesis site and the structure providing the source material.

Chapter 17

Creating Encoded Biological Constructs

With the ability to synthesize peptide chains which can be directly encoded into DNA as introduced in Chapter 16, there is a mechanism available to introduce an enormous content of genetic information to the cell. There is a major difference in the random occurrence of information through genetic means, as now rather it is possible to allow amino acids to self-assemble about relevant structures, and when a stable structure results, the Peptide-DNA Synthesizer-Encoder process is used to encode the resultant into DNA. Thus there is no trial-and-error, and it is possible to allow the amino acids, of which there are twenty, to conform based upon their residues to the site, and define what is worthwhile encoding based upon the stability of the structure. The structures to which the amino acids conform can be structural, chemical, or combinations of the two, and thus it is possible to encode for critical factors such as the cytoskeleton, receptors, enzymes, and much more, as this chapter describes herein. This chapter thus addresses another basic discovery of this research monograph as it identifies the fundamental source for the high information content comprising the cell, including structural, regulatory, and enzymatic function.

17.1 Scaffolds for Encoding Constructs of Structural and Regulatory Function

For protein structures that primarily perform a structural function, such as support or storage, a transportation function, or a regulatory function, such as a transcription factor, the method for its initial encoding is introduced in this subsection. The methodology involves starting with a core structure, from which amino acids self-assemble in a stable configuration. This congregation of amino acids, which are connected via hydrogen bonding, are then strung together through the formation peptide bonds, while at the same time, encoding the self-assembly of amino acids into DNA, such that it can be replicated and propagated forward.

The structures to which the amino acids can self-assemble is referred to as a scaffold. This contains the basic structure, but should have functional groups, or contours to which the amino acid components can associate through hydrogen bonding of its functional groups and residues or demonstrate structural conformity. In Figure 17.1, representative scaffolds are shown as stacks of single and arrays of planar steroids. A column of CG-complementary planar steroids is indicated in Figure 17.1(a), consisting of eight dimers arranged in alignment, stacked together, which therefore has sixteen planar steroids arranged as a strand, which could be useful for framing filaments. Flat sheets as shown in Figure 17.1(b) of side-by-side arrangements are shown, which consists of sixty-four planar steroids. In four columns, hydrogen bonding keeps the dimers together as well as the columns, and the interior of the section of hydrophobic. In Figure 17.1(c), a two by two assembly of planar steroids is shown, a total of sixty-four planar steroids to form a three dimensional construct as a unit cell, which can be expanded. The AT-complementary planar steroid is shown in Figure 17.1(d), which will be useful for configuring amino acid assemblies along the entire structure, rather than mainly the ends as in CG-complementary planar steroids.

Applying the synthesizer-encoder site and example presented in Chapter 16 to the construct of Figure 17.1, the structure with amino acids extending up the end of the assembly is presented in Figure 17.2. This mechanism will continue to feed the synthesizer and encoder if there is a source of amino acids that can self-assemble to the chain of planar steroid molecules. It would be expected that the peptide formed would also conform to the strand from which the amino acids conformed, and thus linear constructs of peptides would be expected to be encoded within the DNA. Upstream to this configuration, although not shown, is the procedure to form the complete DNA strand, as indicated by the methods of Chapter 8.

The feed of the amino acid chain to the site of peptide bond synthesis is indicated in a side view in Figure 17.3. It is seen that the amino acids take on the typical format of a peptide chain in which the residues alternate the side-to-side location with each progressive position of the hydrogen bonded chain. Thus, while it will not be exact, the formation of peptide bond would expect to exhibit similar residue configurations as that of the self-assembled process. In addition, it is seen that the configuration is navigated in that the next amino acid in the queue after the one engaged in the formation of the peptide bond is slightly displaced from the linear chain. However, it is connected through hydrogen bonding to the planar steroid end group and to the amino acid undergoing reaction, and thus would be expected to move onto the

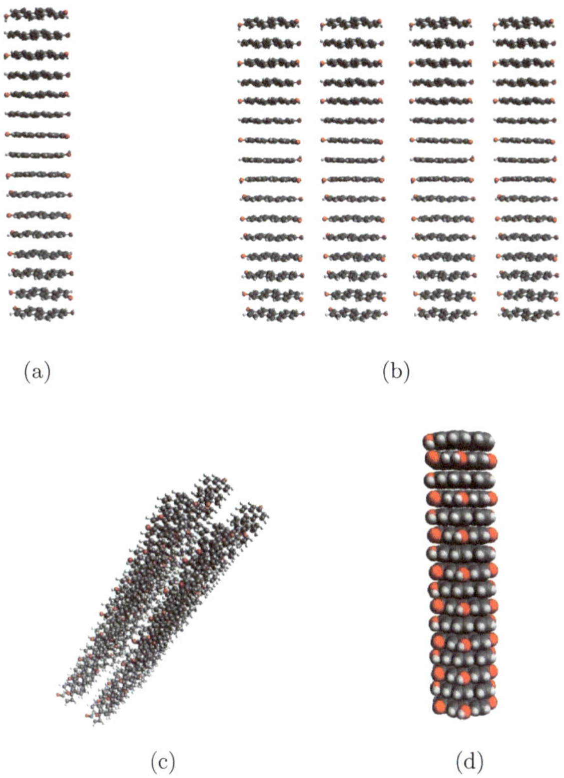

Figure 17.1: For planar steroid arrays, (a) long stacks are possible and (b) side-by-side stacks are feasible.(c) three dimensional constructs (d) with middle and end functional groups as per AT-complementary planar steroids.

reaction site when the peptide bond process is completed and moved one spot as the next-up for encoding.

The distances associated with the progression of the amino acids towards the feed site are indicated in Figure 17.4, which shows extensive hydrogen bonding liking the amino acids to the adjacent units, as well as to the end groups of the planar steroids. It is 1.97 Å between the peptide undergoing bonding and the next one in line, which is projected slightly away in an angled format. The distance is 1.93 Å to the third planar steroid and the amino acid behind the one engaged in the formation of the peptide bond. The distance to the wall then is 1.86 Å hydrogen bond length, and then 1.71 Å for amino acid to amino acid arrangement. The residues alternate in

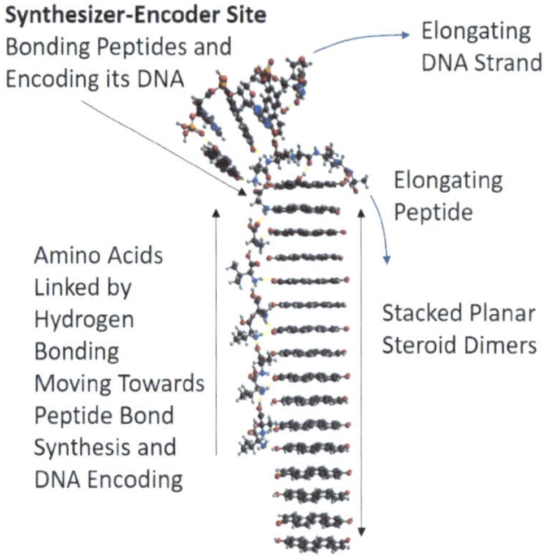

Figure 17.2: A configuration for feeding amino acids to the peptide synthesizing and DNA encoding mechanism. The amino acids conform to the linear structure of amino acids and may be expected to have a linear arrangement after peptide bond formation.

the side position between adjacent amino acids.

The van der Waals representation of the entire assembly indicate an organized configuration, as shown in Figure 17.5. The alternating amino groups along the side can be noted as well as the placement of residues. A significant number of molecules are structured appropriately without conflicts, and the interior hydrophobic core is maintained. The hydrophilic end groups are also noted, as they will drive the flow of amino acids towards the synthesizer-encoder site. The CG-complementary molecules are selected for this demonstration, and thus a mid-molecule functional group is not present, as it would be for AT-complementary molecules, which would enable additional constructs that could surround the planar steroid assembly.

Eventually, I will evaluate the interaction of the peptide-DNA synthesizer-encoder using DNA itself as a scaffold site for the self-assemble of amino acids, but first will analyze the impact that the phosphodiester-sugar chain has on the procedure before considering the double helix format. This arrangement would be the case for the unified complex, where the planar steroid molecules are arranged and bonded, but prior to its nucleotide synthesis and release. In Figure 17.10, this arrangement is shown in which the synthesizer-encoder is attached and the amino acids feed into

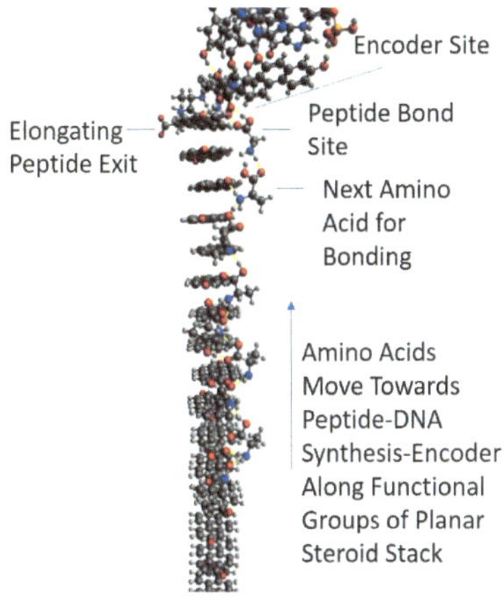

Figure 17.3: The configuration of the amino acids transported through hydrogen bonding to the reaction site. The next amino acid in the queue after the one engaged in the formation of the peptide bond is slightly displaced from the linear chain but is connected through hydrogen bonding to the planar steroid end group and to the amino acid undergoing reaction.

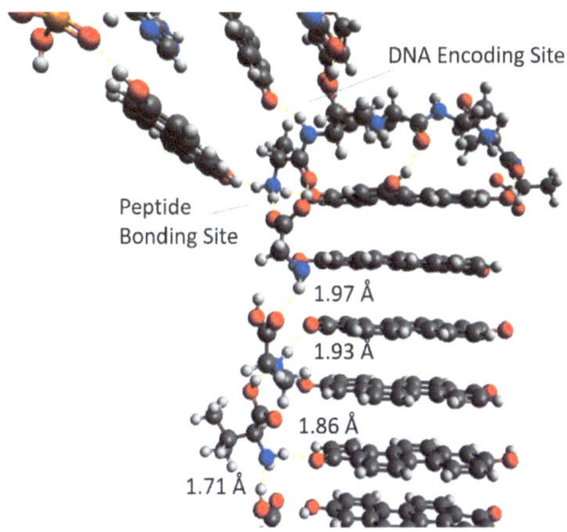

Figure 17.4: The bond distances between the hydrogen element and the functional element, either oxygen or nitrogen, is indicated to show strong strength to permit stable environment and transport method.

the complex. Thus, the unified complex is a scaffold site for the formation of proteins, which will conform to its construction.

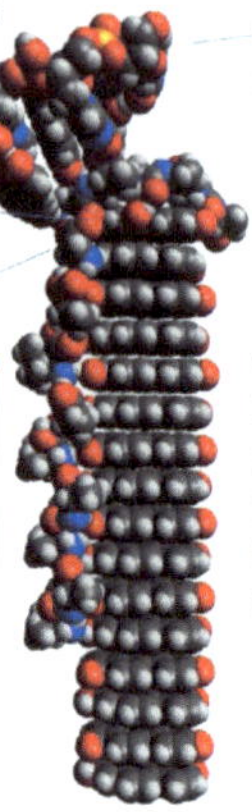

Figure 17.5: The van der Waals configuration for the entire assembly of amino acids feed into the synthesizer-encoder mechanism for peptide bonds and DNA strands. The CG-complementary molecules are selected for this demonstration, and thus a mid-molecule functional group is not present, as it would be for AT-complementary molecules. A hydrophobic core is maintained while hydrophilic ends drive the flow of amino acids.

A different perspective of the impact of the phosphodiester chain on the transportation of the amino acids is indicated In Figure 17.10. It indicates a separation of the amino acid chain from phosphodiester-sugar linkage, in which the amino acids link onto the planar steroid end groups in a regular fashion to move towards the synthesis site. This limits interference of the phosphodiester chain on the integrity of the hydrogen bond coupling the amino acids, and thus makes it feasible for interaction with configurations comprised of planar steroids in conjunction with phosphodiester-sugar linkages.

For this configuration with the phosphodiester-sugar linkage, the van der Waals representation is indicated in Figure 17.10. The representation is still organized even with the linkage as the spacing and placement is consistent with the overall flow and configuration of the synthesizer-encoder assembly. The arrangement of the amino acids along the side chain is similar as well, indicating alternating residue groups as seen in this simulation.

The integrity of the assembly is maintained through hydrogen bonding, which, as indicated in Figure 17.9(a) for the amino acid chain and in Figure 17.9(b) feeding the synthesizer-encoder site. The AT-complementary steroid serving as the catalytic

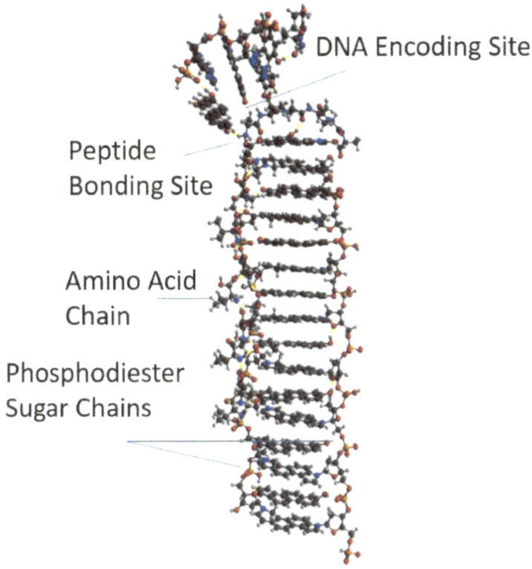

Figure 17.6: For a unified complex in development, prior to the development of the nucleotides but after attachment of the phosphodiester-sugar coupling is indicated, showing that the linear framework is feasible for transporting the amino acids up the chain to the synthesizer-encoder site. Thus, additional configurations are possible and interactions with the bond formation process of the nucleotides is also feasible.

site maintains hydrogen bonding at a hydrogen bonding distance of 1.71 Å with the peptide chain at the mid-molecule functional group for this configuration, indicating that the interference with the phosphodiester chain is minimal. The hydrogen bonding of the peptide bond formation site is 1.65 Å and 1.83 Å for binding of the planar steroids to the lead peptide, and 1.76 AA for the incoming amino acid to be bonded onto the elongating peptide chain.

The DNA molecule as a potential scaffold site is evaluated to determine if it its structure can be utilized in a similar manner as to that of the linear planar steroid chains, with the result of a simulation indicated in Figure 17.10. To produce the DNA molecule, the unified complex structure is fully synthesized, and the planar steroid molecules are ejected to produce the double helix, which reduces the spacing. Consequently, although it is possible to get a few amino acids to transfer towards the reaction site for peptide formation, it is challenging to develop long chains without interfering with the phosphodiester linkage as the amino acid chain needs to run in between the loops.

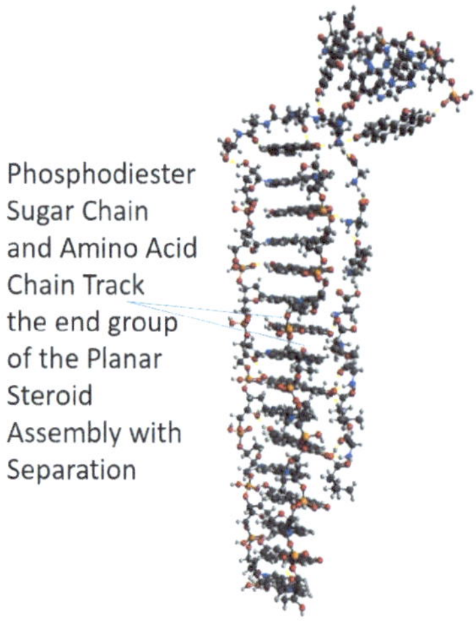

Figure 17.7: Viewed from a different perspective, the amino acid chain is separate from the phosphodiester-sugar chain, linking onto the end groups of the planar steroid molecules that are situated in between the bonded planar steroid. This limits interference of the phosphodiester chain on the integrity of the hydrogen bond coupling the amino acids.

However, it is possible to transfer the amino acid chain conforming to the DNA double helix by association with a linear planar steroid chain located in proximity, as indicated in the arrangement of the molecular simulation of Figure 17.11. Thus, the peptide chain and encoded elongated DNA chain is performed on the planar steroid assembly, even though the majority of the sequence is associated with the conformal map of the amino acid segments on the DNA molecule, in association with the grooves and in the interaction with the nucleobases. Some of the resultant polypeptide is associated with the transition to the remotely located planar steroid assembly and encoding site. As discussed later in the chapter, this is relevant to the formation of transcription factors.

17.2 Scaffolds for Encoding Constructs of Enzymatic Chemical Function

While the prior section focused on scaffolds that produced proteins whose end function would be primarily structural in nature, this section concerns peptide structures

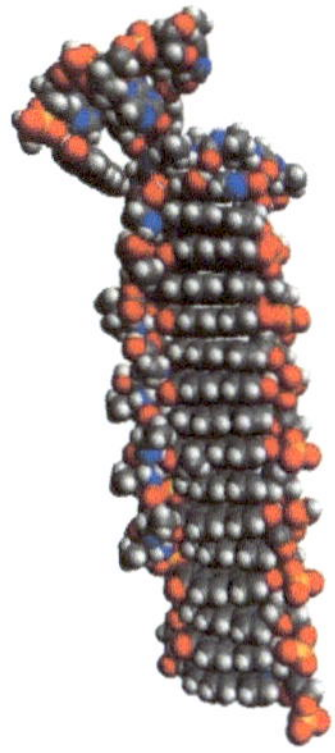

Figure 17.8: The van der Waals representation is indicated for the amino acid, phosphodiester, unified complex configuration. Note that this assumes that the amino acid and planar steroid molecules co-existed, as it will be recalled that the amino acids were generated from the unified complex comprised of planar steroid molecules, and partial synthesis with out-of-phase binding.

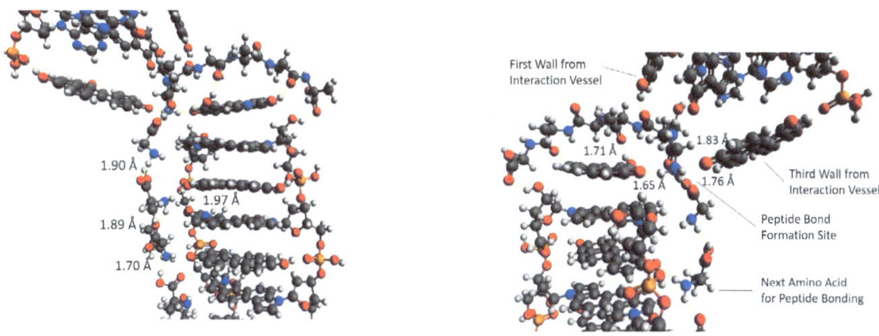

(a) (b)

Figure 17.9: Hydrogen bonding maintains the integrity of the amino acid chain (a) feeding the synthesizer-encoder sites (b). The AT-complementary steroid serving as the catalytic site maintains hydrogen bonding with the peptide chain at the mid-molecule functional group.

whose end function is related to enzymatic activity of a chemical nature resultant, for example, in catalyzing a chemical reaction. This configuration thus requires a

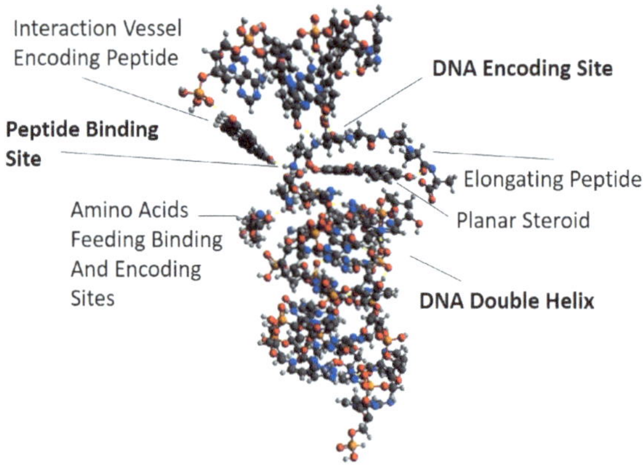

Figure 17.10: A planar steroid at the top site of the double helix is indicated in which amino acids are fed into the synthesizer-encoder sites. The amino acids travel in the internal spacing in between the phosphodiester chains in making its way to the peptide bonding site. It is possible to maintain integrity of the amino acid chain in this transport process, but it is complicated by the close proximity of the phosphodiester chain relative to the hydrogen bonding distances, thereby making integrity of the chain more difficult than the linear chain.

construction of a catalytic site comprised of amino acids with residues consistent with functional transformation. The example considered in this section uses histidine and glutamic acid to make its points.

To produce a scaffold capable of producing zones that form a volume which can contain a ligand interacting with residues, an initial analysis indicates that the out-of-phase binding of the planar steroid assembly with full synthesis prior to strand separation. This region would produce such sites for association of an aggregate of amino acids which can interact with the exposed nucleobases after the non-covalently bond nucleobase diffuses from the site. If the stability is maintained through the association of an accumulation of amino acids, then it can be encoded in a nearby planar steroid construct, similar to that seen with the DNA double helix. Thus, this relation would involve a resultant genetic sequence that produce an enzyme to perform a chemical operation.

A molecular simulation serves as illustrative of the procedure. In Figure 17.12, the procedure to form a scaffold for production of an enzymatic assembly starts with an out-of-phase construct of planar steroids, shown for a portion of six units.

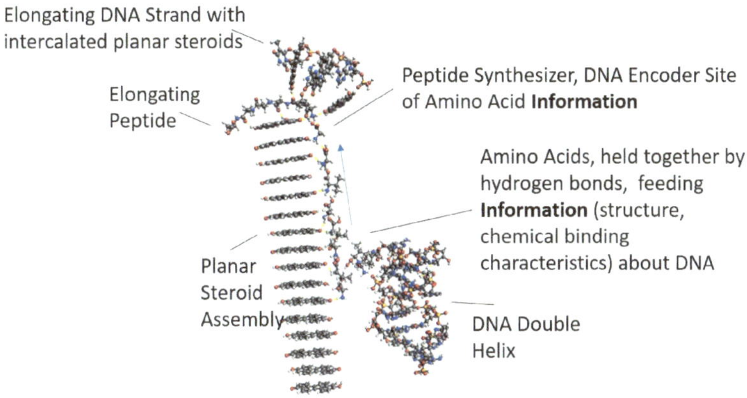

Figure 17.11: Because of the challenges associated with the DNA double helix, and to increase access of the inner nucleobases without having to traverse long distances, a neighboring structure is used to catalyze the peptide chain associated with the DNA double helix, and other molecular structures. The bridging also needs to be considered and will be incorporated into the peptide structure.

This assembly exhibits covalent bonding at only end of the planar steroid to the phosphodiester-sugar linkage, which the other end is free.

After full synthesis of the out-of-phase configuration, the result is indicated in Figure 17.13, which consists of complementary planar steroids bonded to the phosphodiester-sugar chain, along with nucleobases covalently bond to the other phosphodiester-sugar chain. For this example, three AT complementary planar steroids and three adenines would be the result. As seen in the dashed lines, there are confined zones that are available to the external environment since the double helix will not form. The open sites will the areas for interaction either directly to the nucleobases, or by the insertion of a ligand, and the interaction with the ligand. Here, the focus is on the interaction of a pair of nucleobases.

In Figure 17.14, the simulated example is extended to include a chain of amino acids that interact directly with two of the nucleobases. In this case, hydrogen bonding directly between histidine and adenine, and glutamic acid and adenine is indicated. The other amino acids associated with the chain are alanine and two valines. This example can be extended to larger numbers of arrangements of amino acids about the site, and as mentioned ligands can interact with the open spaces and then interact with the amino acids to form a volume of activity about the catalytic

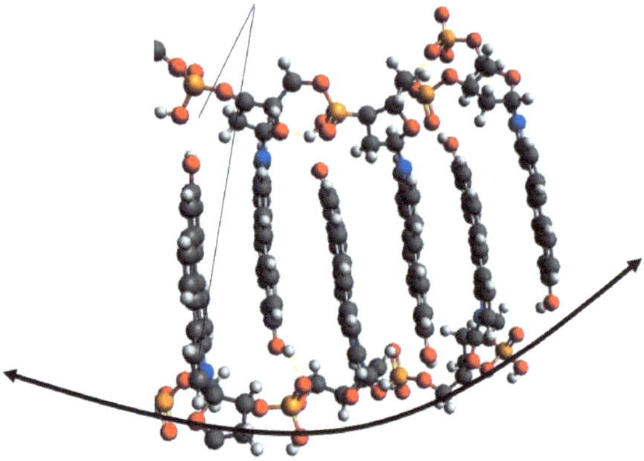

Figure 17.12: To form DNA encoded enzymes, a procedure begins with an out-of-phase binding of an assembly of planar steroids, which covalently bonds one planar steroid to the phosphodiester-sugar linkage, which the other end is free. Here, a portion of the planar steroid assembly is shown.

site.

To complete the simulation, in Figure 17.15 the amino acids that are associated with the open site are encoded into DNA by use of the strategy deployed for copying the structural aspects of DNA, which is to associate nearby a planar steroid assembly that can catalyze the formation of the peptide bond. The amino acids are transported to the catalytic site and then strung together while at the same time undergoing encoding into DNA. Thus, the catalytic site is formed as a prototype peptide and encoded into DNA for genetic purposes.

This capability to produce enzymes is captured in Figure 17.16.The volumetric structure with functional structures are associated with the catalytic residues which can bond either on the functional sites, or on a ligand if it is present and can offer overall system stability. If so, then the amino acids are encoded into DNA. A prototype peptide is also generated for the cell, which will assist in the functional nature of that particular cell and its possible next steps to further the propagation of the cellular function.

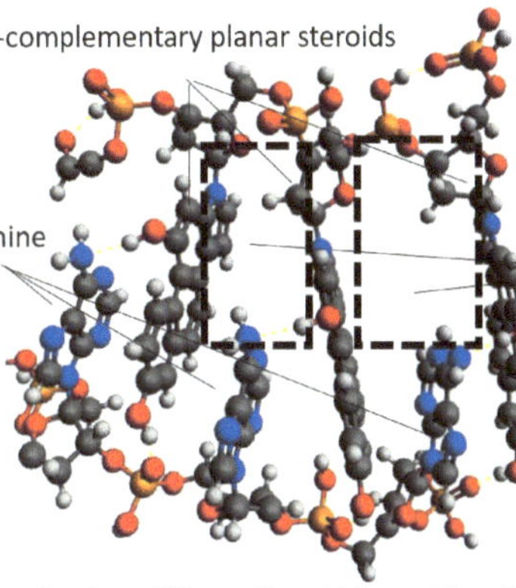

Figure 17.13: After full synthesis, the out-of-phase planar assembly consists of complementary planar steroids bonded to the phosphodiester-sugar chain, along with nucleobases covalently bond to the chain. This example shows three AT complementary planar steroids and three adenines. This configuration creates open sites for reaction with the external environment. This can be filled for example with ligands interacting with nucleobases or can be left open for direct interaction with the nucleobase.

17.3 Information Creation by Encoding DNA from Amino Acid Assemblies

It appears that many constructs were encased with amino acids which were subsequently encoded DNA from the initial generation of information from planar steroid assemblies (the basic construct of a cellular enclosure, DNA, and supporting materials including amino acids and scaffolds). The scaffolds were both of a structural and chemical nature, and thus resulted in encoding for basic structures such as the structural configurations of the cell: cytoskeleton, tubules, filaments, vesicles as well as storage or transporting vesicles. In addition, it encoded for chemical function such as enzymes as chemical catalysts, as well as others that had both structural and chemical binding characteristics such as transcription factors. The overall diagram of Figure 17.17 is indicated for the massive generation of information content resulting after the initial production of planar steroid assemblies and its first products of

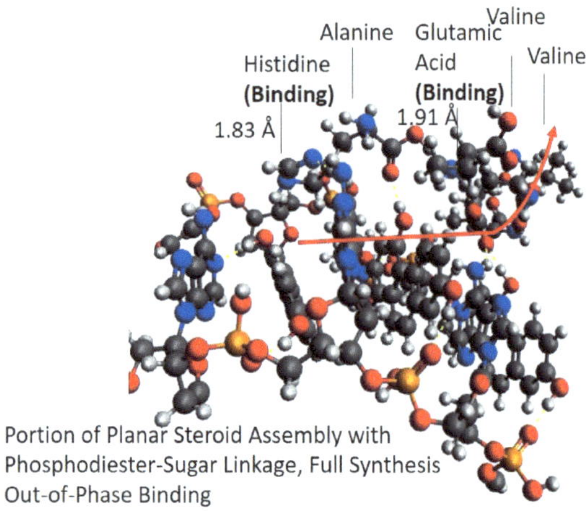

Figure 17.14: An example is shown where amino acids interact with two of the nucleobases, exhibiting hydrogen bonding with adenine of histidine and glutamic acid. The other amino acids are valine and alanine which connect the amino acids together to form a single chain. This example can be extended to larger numbers of arrangements of amino acids about the site. Ligands that can interact with the space can also be inserted for interaction with amino acids.

the cell, DNA, and the supporting material.

As a summary, Table 17.1 lists some initiating scaffolds, which may be an interactive site, such as nucleobase, or a passive structural element such as planar steroid assemblies, which are responsible for the tremendous increase in information content after the initial cellular construct. There are more structures than those listed, but these suffice to demonstrate the point. The scaffolds derive from a mixture of in-phase components, such as with the nucleic acids, and from out-of-phase components. Those resultant peptide structures derived from in-phase components tend to have function associated with transcription, translation, and replication, whereas the peptide structures derived from out-of-phase binding structures tend to have structures associated enzymatic and structural support. Included in this list are proteins as scaffolds themselves for the further production of compatible protein, composing sub-unit assemblies. It is noted that the structure of transcription factors such as the leucine zipper and zinc fingers are consistent with the process described in this chapter of encoding off-site from the DNA double helix via the planar steroid assembly comprising the peptide-DNA synthesizer-encoder.

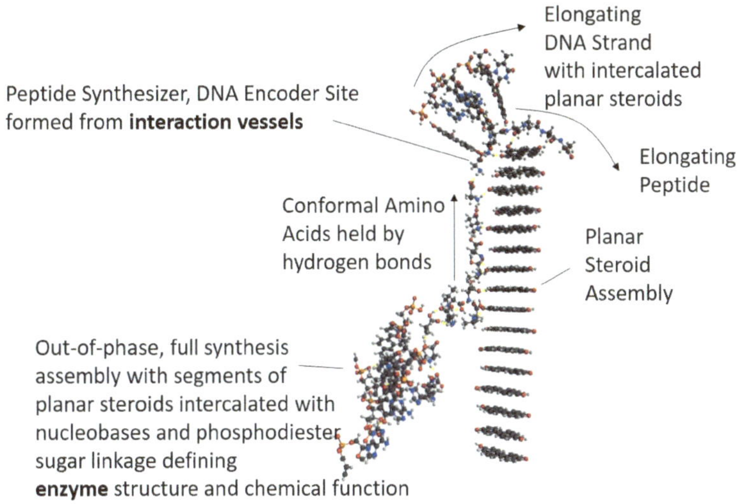

Figure 17.15: These amino acids can then be encoded into DNA by using the inverse protein to DNA conversion mechanism, which here is shown as a neighboring planar steroid assembly. As the amino acids are attracted to the end groups and traverse to the head of the assembly, they will be secured to the AT-complementary planar steroid. With the association of interaction vessels, this permits a peptide bond to be formed. After the peptide bond is formed, it is then encoded into DNA as a trinucleotide codon in accordance with the translation tables developed in Chapter 6 for interaction vessels.

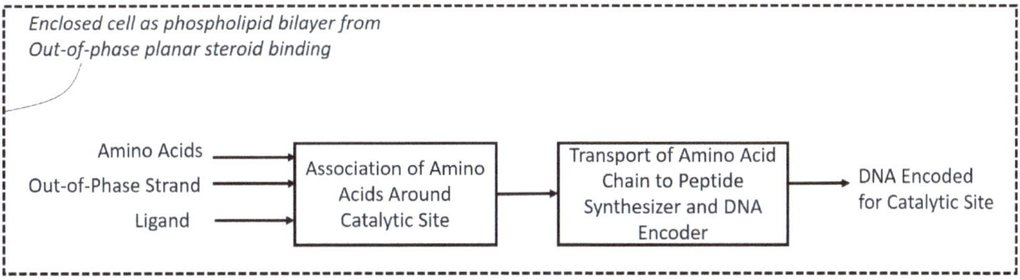

Figure 17.16: The process for forming an encoded chain of amino acids associated with the formation of an encoded DNA strand associated with an enzyme.

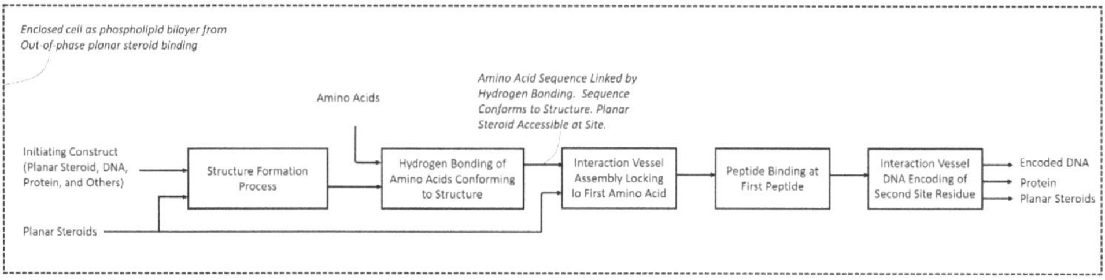

Figure 17.17: The process of creating elongated peptide chains while encoding the chains into DNA so that they can be replicated.

Initiating Scaffold for Information Generation as Encoded DNA	**Resultant Encoded Function**
Stacked Planar Steroids	Microfibrils
DNA	Transcription Factors
Membranes and Vesicles	Protein Globules
Peptide DNA Synthesizer Encoder	Peptide Binding Enzymes
Engaged tRNA, rRNA, mRNA	Ribosomal constructs
Planar Steroids Bound to DNA	Receptors
Dimers	Transport Receptors
Protein Structures	Compatible Protein Sub-Units
Microfibrils	Arrays of Fibers
Single Strand Planar Steroids	Protein Sheets and Helices
Arrays of Fibers	Filaments, Tubules of Cytoskeleton

Table 17.1: Initiating construct and final resultant after amino acid assembly and subsequnet processing through the Peptide-DNA Synthesizer-Encoder.

Chapter 18

Discussion

This manuscript presents a blueprint for life origination and the synthesis of novel genetic systems. The starting material is the planar steroid molecule, $C_{17}H_{10}O_2$, which self-assembles as long stacks and whose ends are complementary for hydrogen bonding as dimers, pairs. The assemblies of planar steroid molecules are arranged in different alignment and binding patterns, and thus provide a rich array of molecular constructs that impart a range of biologic functionality. Moreover, there is one "particular" design that enables a digital molecular communications network resultant as DNA and its complementary planar steroid set that enables subsequent function. Furthermore, the other designs, while not imparting digital content, provide important characteristics necessary to support life function, including the cellular enclosure to enable the storage and transmission of information content.

Foremost, this solution provides a method and means to impart digital information content into molecular structures. Thus, the genetic code has been cracked, deciphered. By arranging aligned dimeric pairs of planar steroid molecules and implementing in-phase binding patterns and interdimeric synthesis strategies, a base four code is the result. The design of DNA is thus feasible. Moreover, an unintended consequence, which in retrospect is a necessity to enable access of the information content provided by DNA, is that the design strategy also induces co-synthesized products of complementary planar steroid molecules that enable the life function operations of transcription, translation, and replication.

The solution for the design of DNA has uncovered the reason behind the heretofore curious mapping of the genetic code involving a trinucleotide encoding for an amino acid. A complex molecular construct with feature sizes on the length scale of 1 nm is responsible for organizing and assembling a sequence of physical vessels, whose structural and chemical properties precisely interact with the residues of amino acids. The stacking property of planar aromatic cyclic structures is an enabler of the configuration, and opportunities for hydrogen bonding is the means to stabilize the structure.

While it was not the intention at the onset, the development of a solution for the design of DNA has also led to an apparent source for the origin of the cellular phospholipid bilayer. By developing a framework for DNA as a digital communications network, of which the constituents represent not only information flow, but also physical molecular structures, a certain clarity results that enables such unforeseen advancements. Analyzing alternative scenarios involving errors due to mismatches in alignment configurations or as potential new features involving alternative binding patterns are rapidly developed with DNA incorporated within the framework of a network topology. In this context, the planar steroid molecule is a physical molecular bit whose tetra-ring structure relatively encodes in base four.

As this manuscript relates to the generation of information content, another important discovery is the source of the massive information content contained in the DNA. It was shown that the planar steroid assembly are capable of constructing the inverse of the current central dogma of molecular biology. That is, to initiate life, it was necessary for information flow to go from the structured peptide assembly to encode DNA. This mechanism is a far more efficient mechanism to produce functional genes which exhibit optimal interaction and association.

In conclusion, a comprehensive solution has been presented for the design of DNA and the genetic code. It is self-consistent and provides conceivable answers to mysterious observations, and even includes a solution for the origin of the phospholipid bilayer cell enclosure for DNA, which is acquired at the same place, at the same time as DNA, and thus, together with the planar steroid molecule set enabling genetic function, provides a solution for the origin of life. It answers the original question I had that drove this line of research: why is cortisol A-T a perfect projective match, which is observable in the DNA of living cells? The developments are extensible as a framework for encoding digital molecular structures, of which alternative and synthetic configurations are envisaged.

Chapter 19

IP-Protected Product Directions

It is correct that discovering an origin for DNA will not only explain the past but also provide a path for the future. In this regard, there are innovations that derive directly from the basic discovery of the structural symmetry and functional relation between steroid molecules and DNA. Work by the author has already begun in analyzing mechanisms of action and novel drugs that can interact directly by binding to nucleic acids, either single strand in a secondary structure or as a double strand [21]. Other aspects of this work involve technology and philosophical advancements, and thus there is a wide array of topics that will be influenced by these results, assuming that a team is effective in delivering the message. Fortunately, the results were anticipated, and several provisional patents were filed along the way of the development path, such that it will be possible to bring in external investment and develop the science and technology in a coordinated way.

19.1 Medical, Health, and Biological

From these results, new advancements in the life sciences, health, and medical related fields are envisaged, for example:

- Synthesis of genetic structures based on assemblies and binding patterns of planar molecules;
- Discovery of biological materials that can be produced from stacked assemblies of planar steroid molecules: For example, one particular design results in DNA, and another design produces the biological cell bilayer membrane, but there are still other designs that need to be explored to determine their outcome, and their relation to life function;

- New life functions and synthetic life forms based on different ring structures, arrangements, and intermolecular hydrogen bonding configurations;
- Integration of systems and cellular models, with new transcription models, for predictive analysis of basic core functions such as body temperature control;
- New understanding of transcription mechanisms and the interaction of intracellular transcription factors as a means to bind symmetric, complementary ligands onto DNA;
- Predictive modeling software and algorithms associated with the efficacy of binding onto thymine;
- Improving the efficacy of synthetic steroids while minimizing or eliminating side effects;
- Novel drugs and supplements based on intermolecular binding directly to the available functional group of uracil in RNA-based diseases or thymine to enhance physiological control;
- New methods of protein transcription and formation through the use of planar steroids as templates;
- Creating biological materials by inverse encoding of DNA from protein structures by use of the interaction vessels;
- Creating new life systems.

19.2 Technology

The advancements described on the design of DNA and genetic codes can enhance many technology endeavors, including molecular scale systems and algorithms, for example:

- Artificial intelligence coupling with the molecular strategies;
- Data storage techniques based on molecular information systems with internal read function;
- Adaptive material synthesis that conform simultaneously structure and function;

- Molecular machines with inherent correction of malfunction;
- New types of molecular constructs that interact with molecules based upon structural and chemical interactions, similar to the interaction vessels enabling prebiotic transcription, translation, and replication;
- Design of computational engines of a molecular configuration.

19.3 Philosophical

Even at the basic thought levels in examining the past and future, the results on the design of DNA and origin of life will enable a scientific basis for discussion as it has aspects of agreement with the central beliefs of several philosophies:

- Search for extraterrestrial life and seeding other planets for life, which will be of interest to national organizations, such as NASA;
- This work should provide a strong and vibrant foundation for a field in molecular structure evolution, which analyze the transition of molecules over time to perform function, which this showed in comparing the transition for example of cortisol from its genesis, the A-T complementary planar steroid. Evolutionary theorists should tend to find these results of utility, and to put the new results into context to answer fundamental questions, which will be of interest to a range of funding organizations and academic institutions;
- The elephant-in-the-room (dinosaur-in-the-room) issue of steroids, and their known effect on physiological size and strength, as well as their effect on personality, will interest evolutionary theorists, especially when combined with the concept of molecular structure evolution from planar steroid assemblies, which will enable new molecular approaches to characterize extinction events based on organism size, for example;
- Placing the results within the context of theological tenets, which would impact the general population to contemplate the meaning of the process of encoding DNA by pairing dimers, the issue of a true master image from which DNA was derived, the duality of the encoded structural relations, and the formation of the double helix, while releasing planar steroid molecules to enable function, thereby resultant in the instantaneous start of life function;
- Interpretation of the results relating to life balance in accordance with the origin of life as a coordinated co-synthesis of form and function.

19.4 Copyrighted Works

As this monograph is comprised of original material, copyright protected works are possible which aim to educate or interpret the results:

- This manuscript as a written description of a solution to the design of DNA and origin of life is a work product that can be distributed itself under copyright protected means;
- Educational books can be augmented with this solution to offer a new understanding of DNA and its relationship with steroid hormones, and to broaden its contents, incorporated with non-copyright protected materials on general topics of molecular and cellular biology, and physiology, using the new design strategy of DNA for differentiation;
- Interpretation of the results in commentary pieces from a series of experts, including philosophy, technology and science, offering insights into the new results.

Appendices

Appendix A

Mechanisms for DNA and Phospholipid Bilayer Synthesis from Planar Steroid Assemblies and Binding Patterns

Starting from the stacked assembly of planar steroids, reaction mechanisms are presented to form the differentiated molecular structures of: (a) adenine-thymine; (b) guanine-cytosine; (c) A-T complementary planar steroids; (d) C-G complementary planar steroids. As a reminder, to encode DNA and its complementary planar steroids, a stacked array of dimers comprised of aligned planar steroids locked at either end by hydrogen bonding. The orientation of opposing rings of adjacent dimers defined the outcome of the synthesis steps, whether the result is A-T, T-A, G-C, or C-G, and whether the planar steroid is A-T complementary or C-G complementary. The appendix indicates the methodology behind the differentiation based on the orientation of the adjacent rings. The separation steps are indicated as well as the final formation of the products. Finally, the common steps are described, including the possible a source for the starting material. There are other synthesis pathways than the ones suggested in this section, however, the overall logical sequence would be similar in terms of the transformation of the molecules into the final products. It is important to note that the synthesis steps occur within the same environment, such that while one base pairing is being synthesized, another aspect may be taking place in parallel, and thus the overall reactive environment needs to be consistent across most steps. At prebiotic conditions and environments, UV illumination will play a significant role and catalyze reactions for amination to take place at appreciable rates via free radical, photochemical reactions [22], as implied in the developments of this section. In addition, as noted in Section A.1, the reactions take place within three dimensional constructs, although some mechanisms are presented in a standard two dimensional framework.

Furthermore, mechanisms are described for the formation of the phospholipid bilayer due to an out-of-phase binding of the same stacked assembly that initiates the DNA synthesis and the genetic code formation, which is produced via in-phase binding. In addition, the source outline is given of amino acids as formed also from out-of-phase binding and partial synthesis.

A.1 Co-enzymatic Coupling

The discovery of a primer molecular shape embedded within the structure of the complementary bases comprising each DNA nucleotide pairing as described in Chapter 2 drives the development of the co-enzymatic analytic framework described herein. The common structure indicates that there is a correlation between the DNA nucleotides and planar steroid molecules as an enzymatic construct of interleaved planar steroid starting molecules, of which one section is covalently bonded to a phosphodiester backbone, while the adjacent structure is connected through hydrogen bonding in close proximity. The result is an arrangement of reaction vessels and co-enzymatic framework for the co-development of sequences of encoded structures, as defined through intermolecular hydrogen bonding, and the relative orientation of the planar steroid structures.

As indicated in Figure A.1, the coordination of the dimers enables a sequence of effective reaction vessels, secured at the ends through phosphodiester covalent coupling. This confined environment enables the potential for an enzymatic capability in which the temporary formation of bonding arrangements to permit intermolecular cyclic structures that can establish element substitution, as well as the degradation of chemical bonds to reduce intramolecular stress, and the differentiation of the templates into their individual capacity as a nucleotide base.

A.1.1 Chain Linkage

In Figure A.2, the method of linking stacks of planar steroid molecules is indicated, in which the enzymatic characteristics of the implicit reaction vessel are implemented. The initiating step involves the amination of what will be the steroid molecules, that is the non-nucleotide molecule. This occurs at both ends of the molecule, in proximity to an intermediate diketone group on one side and a hydroxyl group on the other side, as indicated in Figure A.2(a). Because this planar structure will enable a catalytic activity, the amine group reacts with both of the ketone groups,

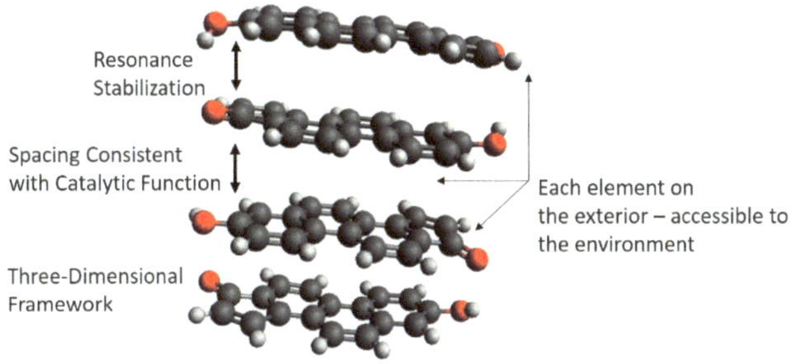

Figure A.1: By interleaving the tetra-ringed aromatic structures, and securing each alternating pair to a phosphodiester backbone, there exists the capacity for an enzymatic function in the development of each molecule.

and the amine group achieves alkene addition on either side of the phenol group, as shown in Figure A.2(b), which then integrates within the DNA nucleotide molecule, displacing the residual as methane and as methanol. This reaction is similar to the fullerene synthesis of [23, 24, 25]. The connection is then achieved by binding onto the inserted amine groups, and then the formation of the sugar group to stabilize the phosphodiester backbone, as indicated in Figures A.2(c) and A.2(d), respectively. Relevant reaction sequences to this synthesis route are the stereoselective Diels-Alder reactions [26, 27, 28], for this type of integrated structure.

This procedure of coupling one of the molecules for each pair results in a construct of a chain of alternating pairings of tetra-ringed structures, secured on both sides by a phosphodiester sugar backbone. This is considered as in-phase bonding. The non-covalently bonded molecule is secured through hydrogen bonding associated with the PO_4H molecule on either side. The spacing of roughly 7 Å between each pair of couplings is consistent with the aromatic stabilization of both molecules and the physical distance of the phosphodiester backbone of DNA.

A.1.2 Differentiation

The implementation of this configuration with positioned oxygen elements involves intermolecular hydrogen bonding to cross-couple the two pairs of dimers, which is critical in defining the differentiation between A-T and C-G pairing. Since the ends are already coupled to connect the pairs, the available oxygen structure of the di-ketone group is arranged through the middle of the molecule, that is the second or third ring. The ketone group is configured two carbons away from the ketone group

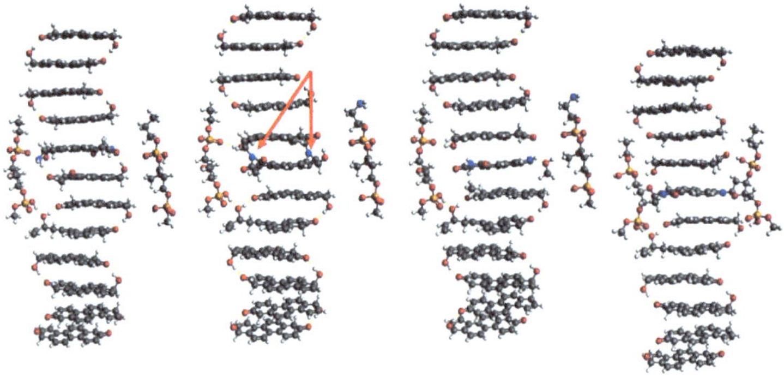

(a) Amination (b) Connect (c) Separate (d) Bond

Figure A.2: (a) To initiate the link of the phosphodiester backbone to the DNA nucleotides, the two amine groups are attached to the molecule, which will become the steroid molecule. (b) a binding is achieved to replace the alpha carbon between the diketone groups, and on the other end, the carbon binding to the hydroxyl group is replaced. (c) after binding, there is tendency for leveling with respect to the steroid molecules above and below the molecule where the amine group was replaced. (d) the final orientation, and ready for further amination and separation.

associated with the end-based pairing. This ketone group is then connected to the ring, or via a hydroxyl group, connected in proximity to the ketone group on the other molecule. The position of the ketone group relative to the adjacent planar steroid molecule of the adjacent dimer will define whether the molecule becomes A-T or G-C. Interestingly, the ketone group associated with the A-T pairing of the steroid molecule will cross-couple to the nucleotide molecule and reduce the chances of amination from occurring, and thus arrive at an adenine structure. This is indicated in Figure A.3(a) and (b) for G-C and A-T respectively, indicating the position. This is demonstrated by looking at the electrostatic potential in Figure A.3(c) and (d). This provides the underlying basis for the differences between A-T and G-C. Here an OH group on the receiving molecule is included, but simply the presence of two oxygen elements on both sides of the molecule, one oxygen affiliated with the hydrogen bonded steroid molecule and the other oxygen associated with the covalent connected molecule, with respect to the phosphodiester chain.

In addition, this issue is also seen on the other side of the molecules, where

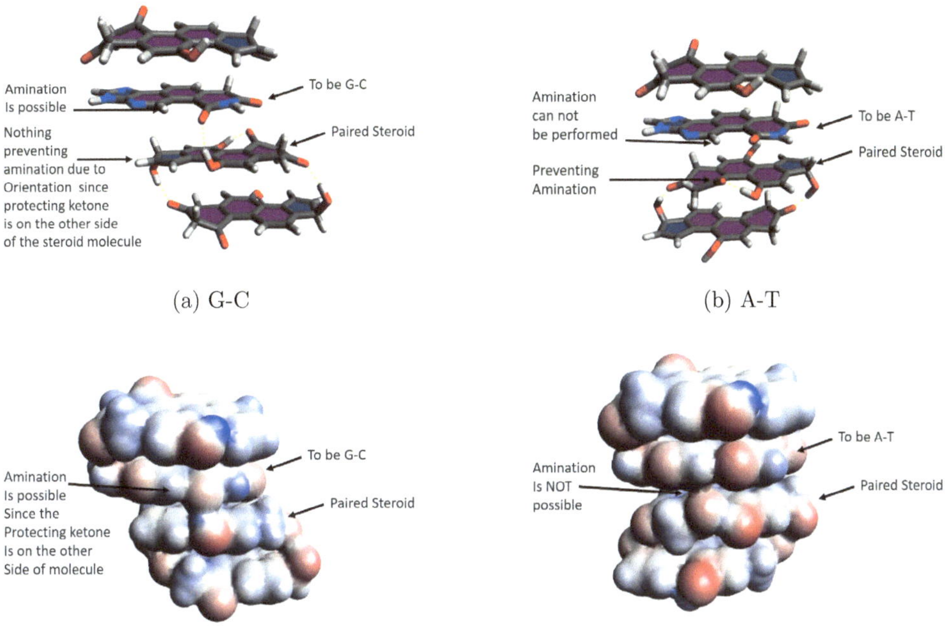

(a) G-C (b) A-T

(c) G-C (d) A-T

Figure A.3: (a) For G-C, the orientation of the planar steroid molecule, second from the bottom, with respect to the molecule that will become a G-C nucleotide is oriented such that amination is possible on the leading element of guanine, as initiated through UV illumination. Thus, an NH_2 group will be affiliated with the leading side of guanine. (b) For A-T, however, there is a connecting group of O-OH associated with the steroid molecule pairing, thus amination is not possible. Ultimately, this provides an explanation for A-T having only two internal hydrogen bonds, while G-C has three. (c) The electrostatic potential indicates that the steroid molecule will not influence the G-C amination. (d) The electrostatic potential indicates the coupling with the steroid molecule which will prevent amination for A-T.

the access of the molecules, and their relationships between the paired relationship of the planar steroid molecule to the nucleotide, which will ultimately become the DNA nucleotide base pairing. In Figure A.4(a) and (b), the other side of the steroid - DNA complex is indicated. The electrostatic potential indicates the interaction of the planar steroid agents and the molecule that will become the DNA nucleotides in

Figure A.4(c) and (d) for the G-C and A-T pairings. In addition, the pairings for C-G and A-T also show similar results, both on the front side and the back side. On the A-T backside, there are no oxygen elements associated with the di-ketone grouping, or ketone-hydroxy grouping, which is more stable, while on the C-G backside, there will be one ketone group associated with the diketone group. This differentiation will enable a differentiation in the subsequent reaction sequences.

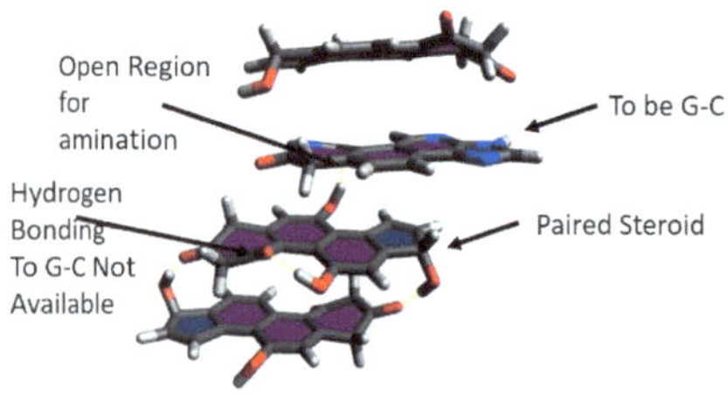

(a) G-C

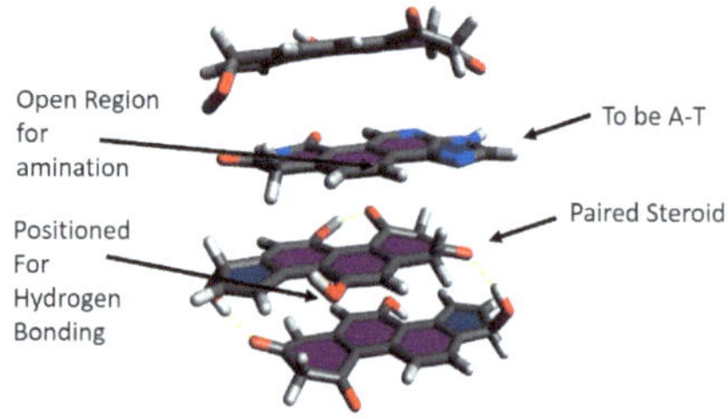

(b) A-T

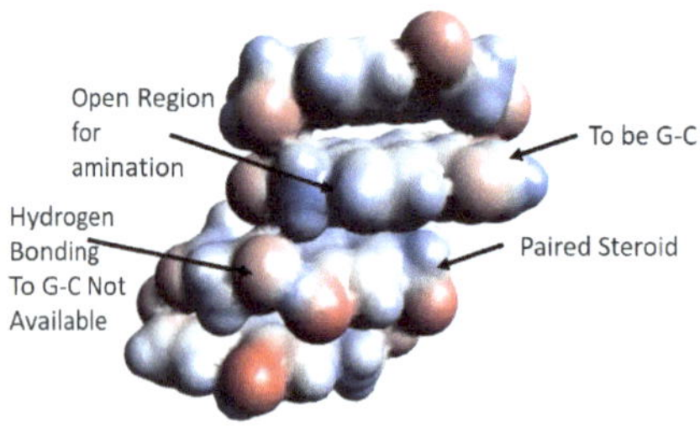

(c) G-C

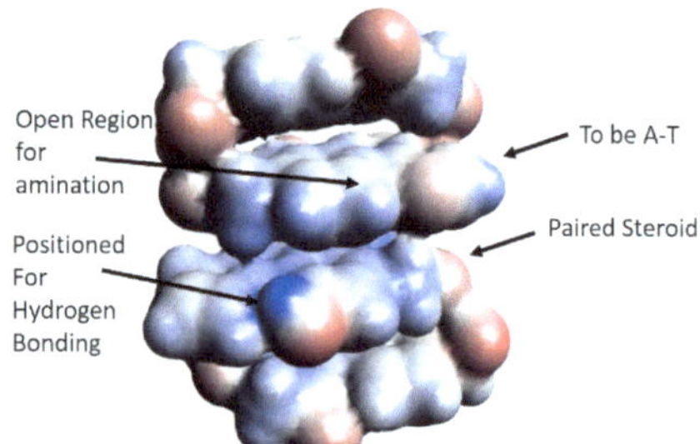

(d) A-T

Figure A.4: On the other side of G-C relative to Figure A.3(a), the orientation of the planar steroid molecule, second from the bottom, with respect to the molecule that will become a G-C nucleotide indicates potential influence over the oxidation and amination. (b) For A-T, there is little association with the backside other than a potential hydrogen bond, and thus oxidation will be effective at the thymine molecule side. (c) The electrostatic potential indicates that the steroid molecule will influence the G-C association on the backside. (d) The electrostatic potential indicates the coupling with the steroid molecule will influence oxidation for A-T.

In binding the phosphodiester backbone prior to its disassociation, as shown in Figure A.5, of which the binding occurs at both ends of the nucleotide molecules 1, 3, 5, 7 and 9, while hydrogen bonding stabilizes the planar steroid molecules 2, 4, 6 and 8. Note that phosphor group contains both OH and an O, such that hydrogen bonding can be achieved at both ends of the steroid molecule: one side of the steroid molecule contains an OH group which will hydrogen bond with the O element of the phosphate group, and on the other side, the steroid molecules comprises a ketone group which will hydrogen bond to the hydroxyl group of the phosphate group. The conversion of the link to a nitrogen element is indicated, and the binding occurs either at the middle carbon located between two ketone groups of the six-carbon ring, which is discussed in the subsequent sections. It is noted that there is sufficient space for occupation of the planar steroid structure in between each DNA nucleotide, as both structures are planar.

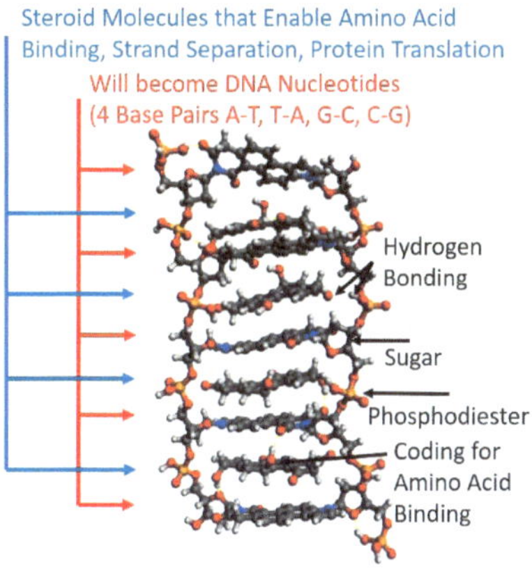

Figure A.5: Model indicates the paired synthesis of the steroid molecules and the DNA nucleotides during a stage of fabrication where the DNA nucleotides are formed, but not yet separated into their basic base pairing with associated hydrogen bonding. There is interaction with the steroid molecule during the formative stages, and the DNA nucleotides influence the formation of the steroid molecule, and the steroid molecule influences the DNA nucleotides. The association with amino acids is not included here.

Continued amination, oxidation, and reduction steps are performed to complete the co-synthesized fabrication of not only the encoded DNA nucleotides, but also the complementary planar steroid molecules. Interestingly, at the end of the synthesis steps, the planar steroids are released from the unified complex when the purine of pyrimidine molecules that are out of position rotate into final configuration compatible with the double helix, that is with the purine molecules back and pyrimidine molecules forward on the 5' side. Thus, there is a culminating step to the synthesis which brings about a life function, as the planar steroid molecules can interact with the DNA double helix to enable replication and transcription.

The overall procedure to produce a differentiated structure is indicated in Figure A.6 for the adenine-thymine pairing. In Figure A.6(a), the differentiation and separation of the DNA nucleotides is seeded by the amination step onto the alpha hydroxy of the steroid molecule, with subsequent attack to the DNA nucleotide in a position to aminate the rings of the purine molecules. Figure A.6(b) shows an amination at the planar steroid and a coupling to the C ring of the DNA nucleotide. In Figure A.6(c), an amination step, and a decoupling of the B ring with oxidation onto thymine, completes the process for A-T. Figure A.6(d) shows the removal of the ethene molecules.

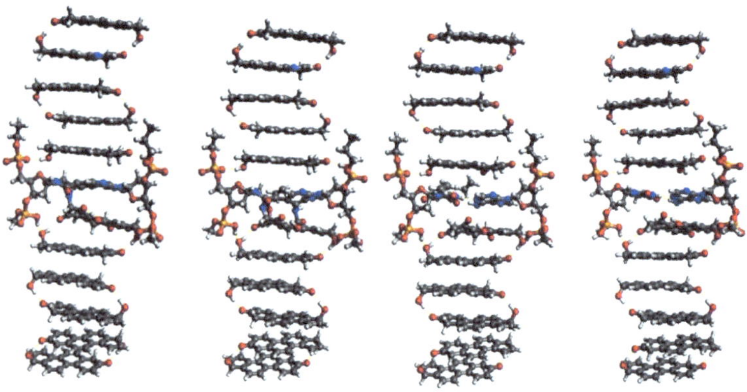

(a) Initial (b) Bonding (c) Aligning (d) Final

Figure A.6: (a) The differentiation and separation of the DNA nucleotides is seeded by the amination onto the alpha hydroxy of the planar steroid molecule, with subsequent attack to the DNA nucleotide in a position to aminate the rings of the purine molecules. (b) An amination at the planar steroid and a coupling to the C ring of the DNA nucleotide, when still intact is achieved to initiate the separation. (c) Completion of the amination, and decoupling of the B ring with oxidation onto thymine completes the process. (d) with the removal of the ethene molecules, there is self-alignment of the pairs to complete the transformation.

A.1.3 Separation

The separation of the planar structure into two nucleobases will follow one of two paths dependent upon the orientation of the ring structure of the paired planar steroid molecule co-catalyzed with the future dinucleotide. As shown for reference in Figure A.7, there are four orientations possible to the intermediate planar steroid relative to the underlying future dinucleotide. Figures A.7(a) and (c) have the ketone group off of the end six-carbon ring oriented downward, while the opposite is the case for the intermediates of Figures A.7(b) and (d). The orientation of the paired future dinucleotide will define the resultant base pairing.

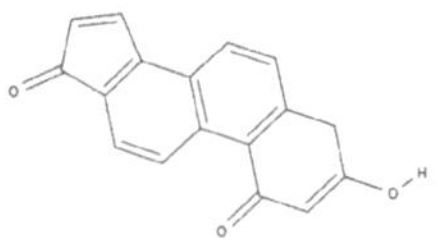

(a) PFBK

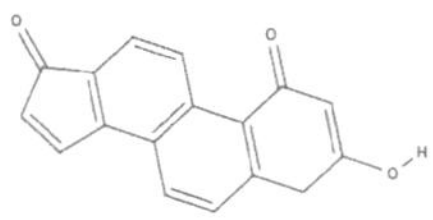

(b) PBFK

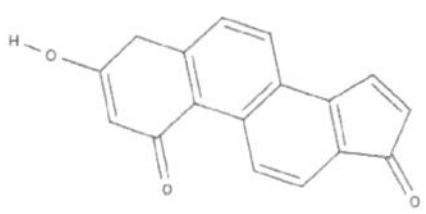

(c) KBFP

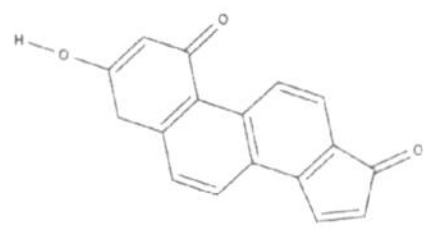

(d) KFBP

Figure A.7: Intermediates of the coupled planar steroid, which has a ketone group in the six-carbon ring that couples into the future adenine nucleotide to thymine pairing depending upon alignment of the ring structures, which will define the capability for hydrogen bonding.

As indicated in Figure A.8, the future dinucleotide will have a common shape and reaction sites, 1, 2, 3 and 4, and the following sequence to define the separation and the nature of the base pairing will depend upon the orientation of the coupled planar steroid intermediate. The reaction mechanism suggested for the formation of adenine-thymine which has hydrogen bonding to the front portion of the adenine-thymine group, which prevents amination from occurring on the adenine front side. This provides a key differentiating step. In addition, on the backside, the interaction leads to the preference for oxidation on the thymine side. Aminations in an ammonia environment, with availability to oxidation components is performed. The entire structure is under stress, in which there is a significant driving force to form the double helix, and thus enable the separation of the base pairs. For the G-C base formation, the suggested reaction mechanism involves an amination on the front of the molecule, at the 3 position, since hydrogen bonding is not present.

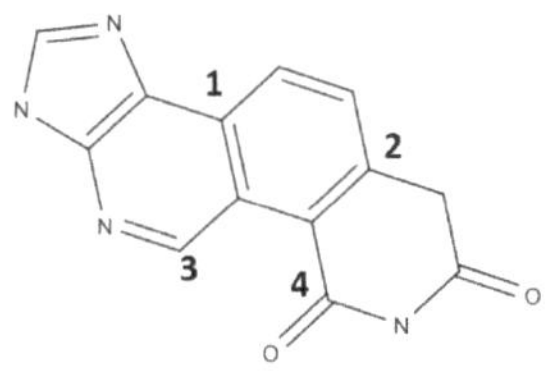

Figure A.8: For the A-T base formation, it is important to recall the result that indicated because of the paired planar steroid molecule positioning of its ketone elements, it is not possible the aminate the front of the molecule, and the amination occurs on the purine side, adenine. This enables the oxidation at the thymine side. The separation then occurs though an amination on both sides of the molecule, with the double bond established on adenine. For the G-C base formation, the result indicated that because of the paired planar steroid molecule positioning of its ketone elements, it is possible to aminate the front of the molecule, and the amination occurs on the purine side, guanine. This orientation also enables the oxidation at the guanine side, and hence amination at the cytosine side. The separation then occurs though an amination on both sides of the molecule, with the double bond established on cytosine.

Thus, this provides the basic answer as to why the A-T pairing only has two internal hydrogen bonds, while the G-C pairing has three internal bonds: because of the relative orientation of it to the paired steroid molecule permits hydrogen bonding. Moreover, it indicates why the overall structure of the base pairings occurs, and provides the driving force for it to occur, high stress, and a significant reduction in energy by transitioning to a double helix form. It will also provide answers to why $NH2_2$ is on the pyrimidine purine side for G-C but on the purine side for A-T, and other such comparative matters. It is important to note that these reactions are effectively taking place over the entire molecule, so the same process steps have to be addressed concurrently on how they would react to A-T if working on C-G, for example.

Subsequently, the steps to produce the pyrimidine products are performed in a reduction and methylation environment. Since the oxygen group of thymine is not engaged in hydrogen bonding and has an adjacent available carbon segment, this will allow for its reduction and subsequent methylation. A reduction of the intermediate product of cytosine will alter the oxygen functional group of what will become cytosine. The positioning of the oxygen group may also enable its reduction, whereas the other segments are engaged with the guanine base, and thus may not undergo reduction. There is literature available that has studied methylation properties of thymine and cytosine from which the synthesis steps may be obtained.

As a final product, the planar steroid molecules appear to have catalytic properties since half the planar steroid molecules, those that are C-G complementary, are unaltered upon release from the unified complex during the formation of the double helix, despite participating and influencing the formation of the dinucleotide. The A-T complementary planar steroid molecules have a single modification, albeit critical, in which a hydroxyl group is added to the C-ring, which enables hydrogen bonding with thymine. For the intermediate products of Figure A.7(a) - (d), the corresponding resultant planar steroid is indicated in Figure A.9(a) - (d), respectively. The ketone on the A-ring shifts to a hydroxyl on the C-ring to maintain planarity, while coupling into the thymine ketone group. Without interaction to protect the ketone group of the A-ring, a hydroxyl group is not formed on the C-ring, and the C-G complementary planar steroid results, which is equivalent to the starting material of Figure 3.3(a).

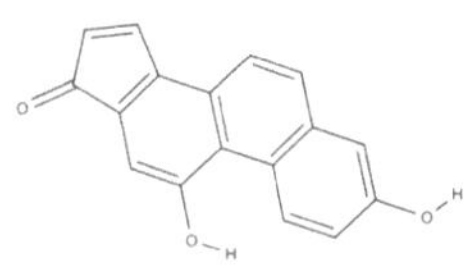

(a) PFBK

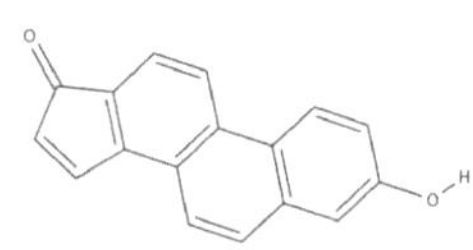

(b) PBFK

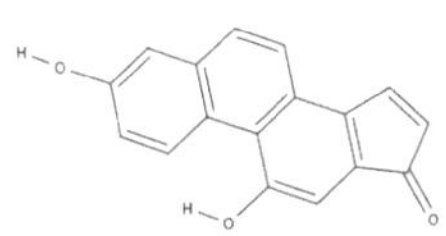

(c) KBFP

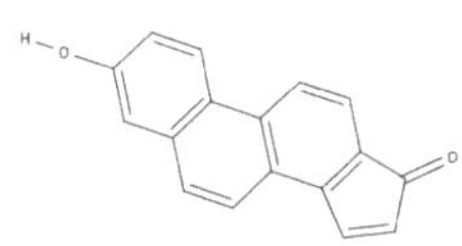

(d) KFBP

Figure A.9: The resultant planar steroid molecules that correspond to the intermediate forms of Figure A.7, indicating hydrogen bonding, thus A-T complementary, or no hydrogen bonding, thus C-G complementary.

A.1.4 Source Materials

Sources for the planar steroid molecule of Figure 3.1(a) are considered. One related structured is the fused aromatic four ring molecule of chrysene indicated in Figure A.10 which is a relatively simple organic structure. Chrysene is a relatively common material, especially in astrobiology analysis, as it is among the class of polycyclic aromatic hydrocarbons. As non-polar molecules, polycyclic aromatic hydrocarbons derive from the thermal decomposition of organic materials and are found in tar and carbon deposits. The molecules are commonly found within the universe and are often associated with the formation of new stars.

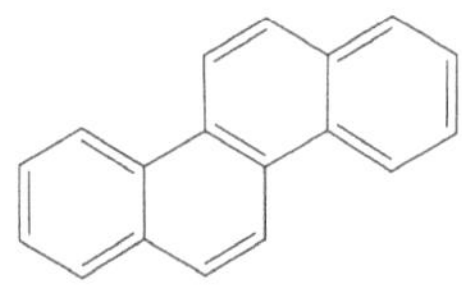

Figure A.10: The similarity of chrysene and the planar steroid molecule is of note.

Chrysene has a similar structure to that of the planar steroid, and the objective then is to convert one end to five-carbon ring and the other end to a phenol. Fortunately, to enable this conversion, chrysene will follow Clar's rule which indicates the positioning of the Kekule resonance structures with the largest number of disjoint aromatic nsextets for polycyclic aromatic hydrocarbons, as presented in Figure A.11.

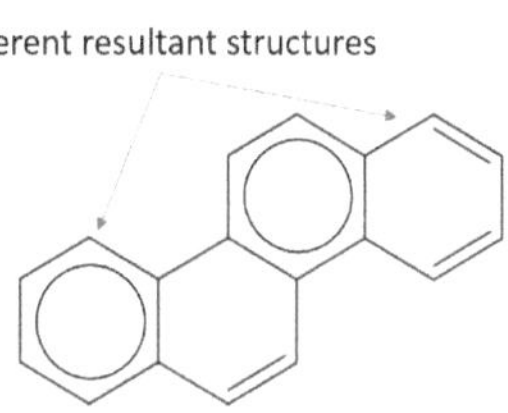

Figure A.11: Aromatic structures of chrysene according to Clar's rule, which will enable different structures on the ends, inducing hydrogen bonding in pairs, and alignment in the central portion, inducing hydrophobic similarity.

Thus, synthesis steps can be considered in converting one end in a differential way to that of the other end. For example, the phenol group can be obtained through several routes including one involving chlorination with sunlight, and a subsequent conversion with sodium hydroxide to convert one end into the phenol group and salt. The five-carbon ring containing the ketone element, can be obtained by the methods described for the conversion of the planar steroid molecule into the nucleobase through reactions in a confined enzymatic framework induced by its self-assembled stacking property.

According to the organic chemist, also an expert on DNA, who I showed a draft of this manuscript and the response to a question set, another potential source for the planar steroid starting material of Figure 3.1(a) are scratches of graphene. Decomposition of a two-dimensional sheet of graphene into an array of shapes may provide the ring structure necessary to synthesize the starting material. Graphene has double bond characteristics and induced defects may lead to the starting material.

In Figure A.12(a) and (b), two alternative starting materials are shown, relative to the starting material of Figure 3.1(a). It makes the DNA synthesis steps simpler, but complicates other factors in which the all carbon arrangement of the steroid molecule is useful in developing molecular constructs. It seems unlikely that these materials would serve as starting materials, but are presented to consider variations of a theme.

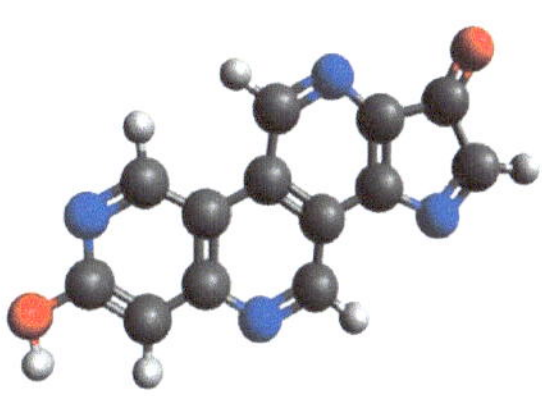

(a)

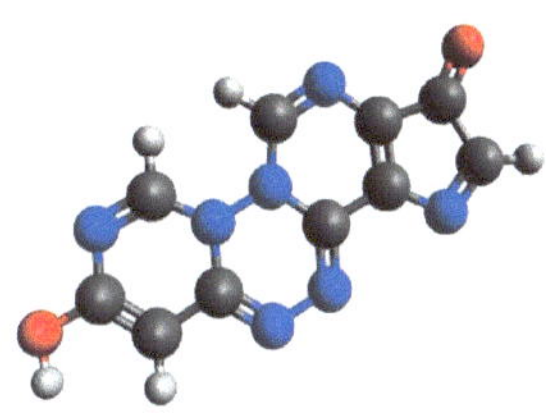

(b)

Figure A.12: Alternative starting materials that have partial planarity and nitrogens incorporated into the carbon array of molecules. These make the synthesis steps simpler.

A.1.5 Forming a Cellular Phospholipid Bilayer

As indicated, an out-of-phase binding of the planar steroid assembly will result in a phospholipid bilayer. The steps to achieve this product is discussed in this section. In Figure A.13(a), the phosphodiester connection at one side of the planar steroid molecules for the situation in which an out-of-phase connection enables the pull away of a single strand away from the planar steroid molecules prior to their complete synthesis into DNA nucleotides. This creates a situation in which the fluidity of the chain is improved by the separation of the chain and a transition to match two planar steroid molecules in dimeric form, as indicated in Figure A.13(b).

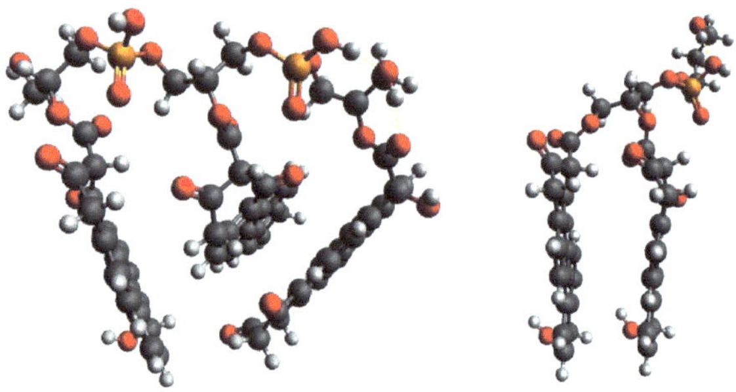

(a) (b)

Figure A.13: (a) The phosphodiester connection of the steroid molecules in an out-of-phase configuration allowing the separation. (b) The rearrangement and separation of the phosphodiester backbone into individual phosphate groups consistent with the phospholipid bilayer.

Through a binding of an amine group onto the hydrophilic section of the phosphate group, as shown in Figure A.14(a), the hydrogen bond onto the chain is prevented, and the group will face the hydrophilic environment. As indicated in Figure A.14(b), this configuration improves the hydrophobic section of the phospholipid structure.

The reduction of the double bond may take place via procedures that reduce the number of double bonds of the planar steroid molecule, involving the break of four bonds to achieve a linear chain, as indicated in Figure A.15(a). The extension of the linear chains results in the phospholipid bilayer molecule structure indicated in Figure A.15(b).

A.1.6 Amino Acid Source Outline

In Figure A.16, the five amino acids of the first row of the translation table are identified in terms of its source due to the partial synthesis of the DNA nucleotide process, but for the situation where the phosphodiester backbone is bound to the

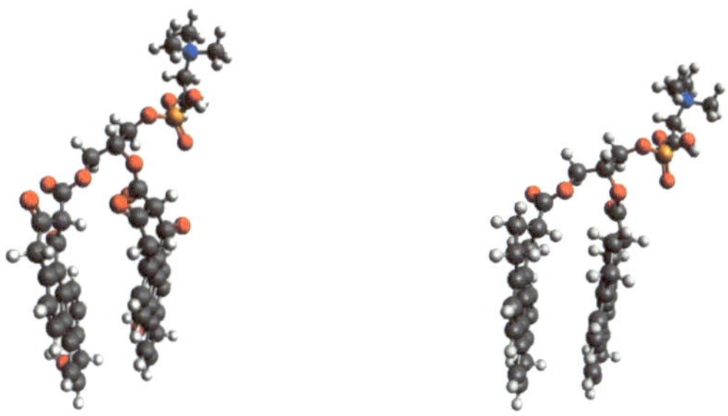

(a) (b)

Figure A.14: (a) Replacement of the hydrophilic head group from the oxygen group to the amine group which will prevent hydrogen bonding to the phosphate group. (b) Elimination of the oxygen groups improves the hydrophobic nature of the phospholipid bilayer structure.

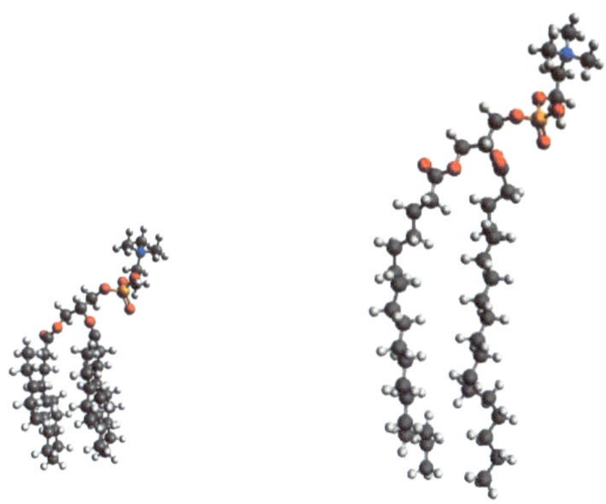

(a) (b)

Figure A.15: (a) Enzymatic reduction of the steroid molecule, and breaking of the bonds to transform to a linear configuration. (b) The extension of the linear chain to the configuration permitting a phospholipid bilayer structure and the biological cell.

planar steroid molecules in an out-of-phase manner. As a reminder, the term out-of-phase means that the covalent bond is formed on alternating pairs on either side of the DNA strands, and was the method used to produce the phospholipid bilayer. Hence, stabilization does not occur, and the synthesis steps cannot proceed to conclusion. In analyzing the individual results, the issue of cysteine involves a sulfur group, which is not inherent to the co-enzymatic complex, and thus is likely the result of a secondary process.

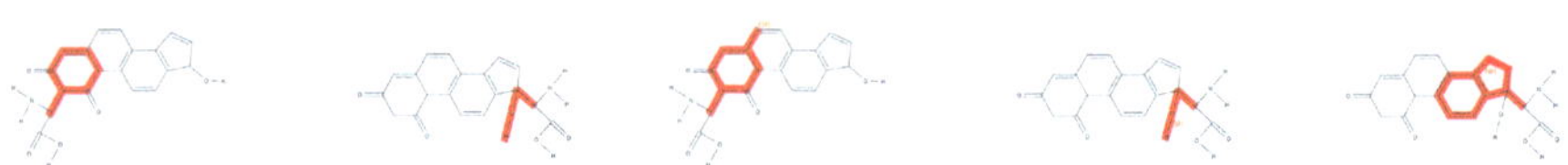

(a) Phenylalanine (b) Serine (c) Tyrosine (d) Cysteine (e) Tyroptophan

Figure A.16: Identifying the source material for the amino acids of the first row.

The second row is comprised of five unique amino acids, and the source materials are presented in Figure A.17. Proline involves binding onto an intermolecular molecule is positioned for interaction with a purine but in this case fuses a ring, five-carbon. Arginine is perfectly represented by the guanine complex which involves three amine groups centered about a carbon and connected by a nitrogenous connection to a two carbon link. Both Histidine and Glutamine are complex configurations, which involve secondary reactions.

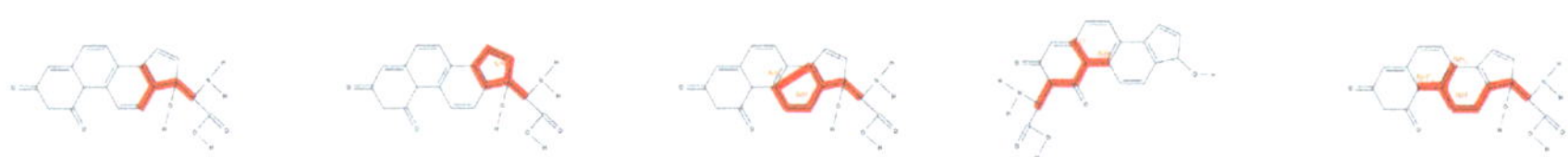

(a) Leucine (b) Proline (c) Histidine (d) Glutamine (e) Arganine

Figure A.17: Identifying the source material for the unique amino acids of the second row.

The third row is indicated in Figure A.18. Threonine is similar to serine and would just involve an additional methyl group, that can be sourced from the purine segment. Again, there is methionine, which has sulfur, and this would require an exogenous input source. Threonine and isoleucine are similar to its counterparts of Serine and Leucine.

For the amino acids of the fourth row, the source material is identified in Figure A.19. For the small residues of valine, alanine, and glycine, both sides will suffice as a start site as either a purine or pyrimidine. The aspartic acid and glutamic acid undergo secondary reactions to achieve the proper result.

With the availability of amino acids, the planar steroid assembly thereby has the key elements to sustain and grow life function: DNA and planar steroid molecules to

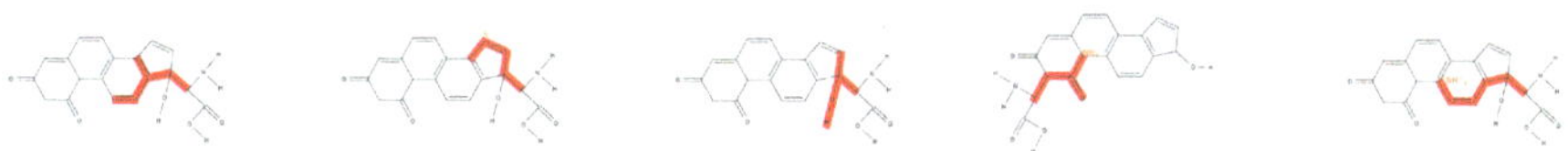

(a) Isoleucine (b) Methionine (c) Threonine (d) Aspargine (e) Lysine

Figure A.18: Identifying the source material for the amino acids of the third row.

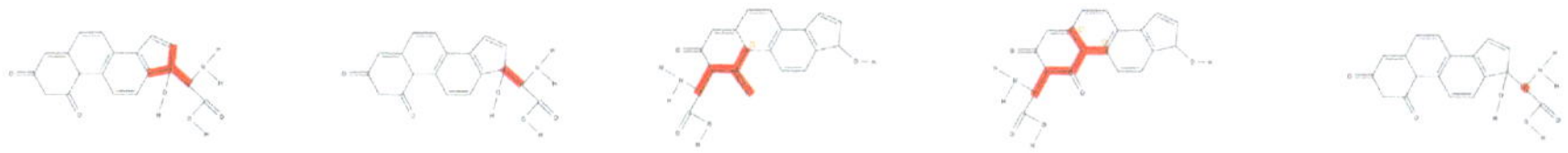

(a) Valine (b) Alanine (c) Aspartic Acid (d) Glutamic Acid (e) Glycine

Figure A.19: Identifying the source material for the amino acids of the fourth row.

induce function, a phospholipid bilayer to enclose the components, and amino acids for growth and regulation.

A.1.7 Alternative Mechanisms

The general strategies developed for mechanisms have alternative pathways, but the general plan would remain the same. For example, there is the possibility that the planar steroid molecule started out as six-carbon on all four rings, with the reduction to a five-carbon group on the DNA nucleotide done during the attachment to the phosphodiester backbone. Other sources for the starting material can be envisaged as well, besides chrysene, there can be derivatives of graphene, or built-up from aromatic rings, including benzene. However, the basic scheme for the communications network would remain unchanged.

Other configurations, which are similar in scope, can be developed, such as with hydrogen bonding within the middle section of the planar steroid molecule via projections made along the hydrophobic interior edges. The main concept would still remain unchanged however, which is that the orientation of the tetra-rings is the overriding and exclusive factor defining the differentiated capacity of the intermolecular pairings that ultimately result in either an A-T or G-C base organization and the base four code.

Still other mechanism strategies may be presented such as using RNA in combination with the phospholipid bilayer model that I developed and described earlier for an enclosure, and then using the planar steroid molecules to interact with the RNA molecules to induce error correction, and double helix formation. However, that approach would involve bringing outside material into the cell, thus there is the issue of achieving sufficient concentration simultaneously for a multi-pot system, and it would not answer the fundamental question as to why DNA has a structure that is consistent with the molecular primer, the planar steroid molecule. Thus, the alternative of combining the planar steroid usage in an RNA world, would seem unlikely.

In Chapter 10, different binding patterns than the primary patterns of the main text are evaluated involving the arrangements of planar steroid molecules.

A.2 DNA Strand Separation

While the DNA nucleotides are in a stable double helix configuration, in order for it to have purpose, it needs to be accessible for reading and action. While the double helix is very stable, it also limits its availability for opening and reading, and without that capability, DNA is just an interesting polymeric molecule, but would serve little purpose. Fortunately, it was co-synthesized from its key, the planar steroid molecule, which can open it such that transcription, replication, and translation can take place. Therefore, in the framework of a communications network, it is necessary to transmit not only the DNA molecule but also its key to open it. Therefore, it is favorable for DNA, and its paired planar steroid molecules, to be coupled in a unified complex within a cellular format to contain all molecules necessary for its function.

Having produced the paired and encoded structures of both DNA, the complementary planar steroid molecules bind to DNA to induce strand separation, which I originally described in [17]. This occurs through energy stabilization of the strands, which are vibrating and oscillating molecular constructs comprising DNA, which is known to have natural modes of oscillation [29]. There is kinetic energy, which comes from the vibratory nature of the DNA strands, and this energy is transferred to potential energy through the stabilization process of binding the planar steroid molecule to the strands. The overall energy function must be balanced and thus the external reduction of internal energy must be dissipated by providing that transferring the stabilization energy into vibratory motion surrounding the stabilization site. This can happen through molecular disorganization at neighboring sties, for example, in this case, by breaking hydrogen bonds, thereby inducing a transcription bubble. An approximate mathematical representation for the process is indicated in Equations A.1 and A.2, for the energy dissipation, $E_s(x)$ about the planar steroid-DNA binding site, via strand separation from its nominal position, represented by $S(x)$. The equations characterizing the strand separation due to planar steroid binding are given by:

$$E_s(x) = \frac{1}{2}(1 + \tanh(k_x x)) \quad \text{(A.1)}$$
$$S(x) = k_b(1 - (\tanh(k_x x)^2)) + S_n \quad \text{(A.2)}$$

where the parameters are associated with hydrogen bond energy $k_x = 1/40$, coupling k_b=3, and nominal separation S_n=1.8 Å, and x runs from 0 to 200 nucleotides, with the result mirrored from -200 to 0 nucleotides about the binding point. These equations can be used to approximate the strand opening away from the binding site. It is dependent upon the energy associated with the hydrogen bonding to keep the strands intact, as when the term $\tanh(k_x x)$ approaches unity, the strand separation will be equal to its nominal value. Other formulations for strand separation can use wave equations to describe the motion of the DNA strands, which will be constrained by the cross-coupling induced by the binding of the planar steroid molecule, and thus would result in reflected wave patterns that will induce interference patterns, and hydrogen bond reduction, strand separation, and bubble formation.

As the DNA molecule is stabilized by binding with the planar steroid molecule, there must be a balance in energy in terms of a release through the disorganization of the non-covalent weakest link, that is the breaking of hydrogen bonds such that the overall energy is nearly equivalent to that of the stabilization provided by the binding event. Essentially, the DNA strands act as two strings in constant motion and with the cross-coupling induced by the planar steroid molecule, that is the triggering element, it acts as a pinch-point, and the DNA strands, which were in relatively independent motion become constrained at one point, and around that point, a small bubble must form through hydrogen bond breaking upstream and downstream of the pinch-point. In Figure A.20(a) the binding of the planar steroid molecule onto the formed DNA complex is indicated in a molecular model as to establish its feasibility. In Figure A.20(b) the electrostatic potential is indicated of the planar steroid molecule to show that the negative potential of the phosphodiester zone will enable alignment of the steroid molecule for binding. In Figure A.20(c), the profile of the strand separation is approximated on either side of the binding site.

It is possible to deploy multiple planar steroid molecules to induce a stronger association and knot of the DNA strands, and thereby improve the duration of the subsequent opening of the strands of DNA. Thus, with intermolecular binding the strongest, it would favor chains of intermolecular coupling TA to form stabilization sites, or longer chains of the weaker CG interactions.

It is noted that there is a bijective code associated with the orientation of the planar steroid type to the location of the bonding point, as indicated in Table A.1.

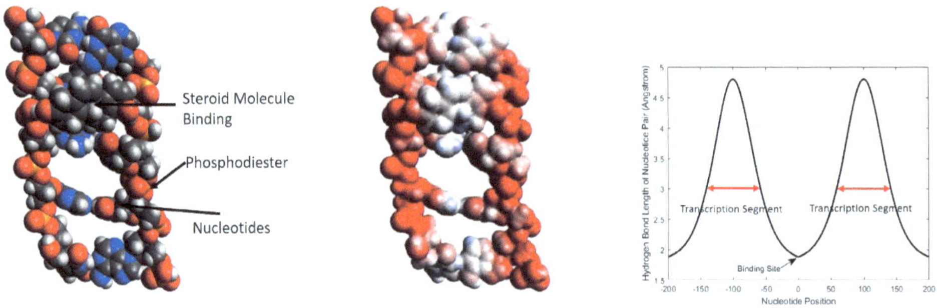

(a) van der Waals spheres (b) Electrostatic Potential (c) Strand Separation

Figure A.20: (a) The binding of the steroid molecule onto the formed DNA complex to induce strand separation. (b) The electrostatic potential of the binding site, which is achieved through hydrogen bonding at both ends, attaching to the phosphodiester coupling. (c) The formation of the transcription bubble on either side of the binding point up to two hundred nucleotides. The quality of the binding point will determine the width of the transcribing separation of strands.

This was originally described in my preprint [19] and indicates that a nucleotide pair and bonding orientation can be associated uniquely for each of the dinucleotides. This is useful for determining the orientation of the planar steroid to complementary base pair to examine transcription scenarios.

Base Pair	**Steroid Orientation**	**Steroid Class**
A–T	$\mathcal{F}$	$\mathbb{H}$
T–A	$\mathcal{B}$	$\mathbb{H}$
C–G	$\mathcal{B}$	$\mathbb{S}$
G–C	$\mathcal{F}$	$\mathbb{S}$

Table A.1: The location and class of the binding planar steroid to DNA nucleotide pair. $\mathcal{F}$ denotes front, $\mathcal{B}$ denotes back, $\mathbb{H}$ denotes A-T complementary with a hydroxyl group on the C-ring, cortisol-like, and $\mathbb{S}$ denotes C-G complementary, testosterone-like.

A.3 Encoded Structures

Further discussion of the codes available in the structure is presented by example. Consider the ten stack of planar steroid molecules of Figure A.21. For the representation, the presence of an oxygen atom is indicated in red.

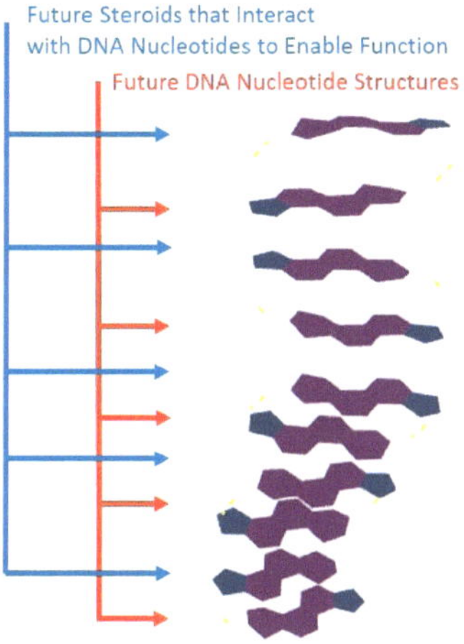

(a) Ring Representation

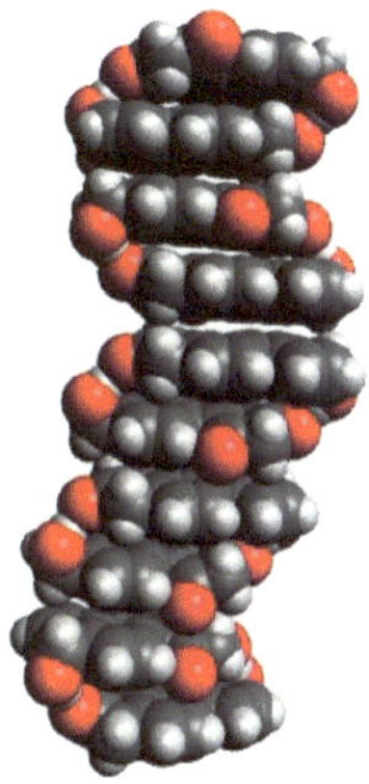

(b) Co-enzymatic complex

Figure A.21: (a) A co-enzymatic structure resulting in the DNA nucleotides and steroids that interact with the DNA nucleotides to enable function after synthesis. It will also include four steroids configured as corticosteroids like cortisol, for molecules 2 and 6, and like testosterone from molecules 4 and 8. The tenth molecule would interact with the eleventh molecule. (b) The van der Waals spheres indicate hydrogen bonding along two edges. Note the code along the middle of the middle of the molecules represented as ketone molecules, in this case from the bottom 0110100101. The other independent base 2 code comes from the edge of the molecule, which enables a base 4 structure.

A representative molecule is indicated in Figure A.22(a), which would be an intermediate form during the synthesis process, which has a ketone group within the six-carbon ring, and its side view is indicated in Figure A.22(b). These molecules are useful for this analysis, although it is the ring orientation that fundamentally characterizes the encoding sequence. But the oxygen atom provides a visible guide for demonstrating the coding principle.

In Figure A.23, the presence ('1') or absence ('0') of an oxygen atom is indicated

(a) Example Planar Steroid

(b) Side View

Figure A.22: (a) A representative material of that shown in the van der Waals sphere representation of Figure A.21(b), which has the ketone molecule off of the six-carbon ring. (b) side view.

towards the 'middle' of each molecule (away from the most distant oxygen atoms on either end). Thus, in reading up starting from the bottom, the first molecule does not have a red sphere representing oxygen, hence it is '0'. The second molecule does have a red sphere representing oxygen, hence it is '1'. Thus, in answering the question, this procedure is continued to produce 0110100101 for the ten molecules. This sequence is used to determine the status of the intermolecular coupling between opposing faces of the dimeric assembly. If there is interdimeric hydrogen bonding, then the molecule representing the DNA nucleotide will either be among the set (A, T), if it has hydrogen bonding, or (C, G) if it does not have hydrogen bonding.

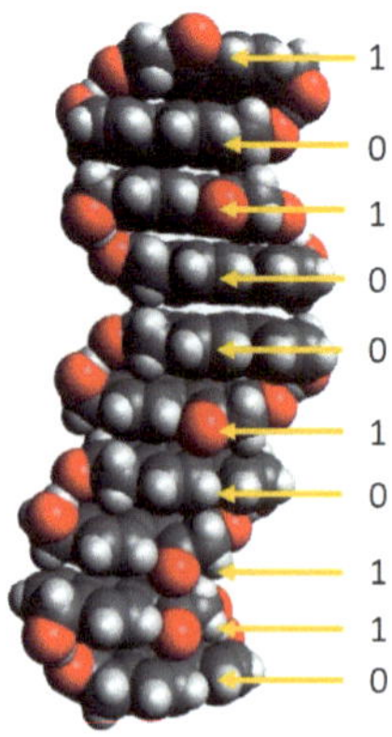

Figure A.23: The coding of the middle oxygen element position of Figure 4.1 read from the bottom is 0110100101 where "0" denotes a hydrogen element in the middle, that is no oxygen, and "1" denotes an oxygen element present.

To further evaluate the code, this example will be continued. It was mentioned in the caption that the other base two code is contained along the edge, which together with the code along the middle of the molecule defines the base 4 structure. For this example, the code along the left edge is shown in Figure A.24, in which '1' represents a six-carbon ring as the leftmost ring, and '0' represents a five-carbon ring as the leftmost ring. For this example, numbering starting from the bottom, the code is 1001011001. If the DNA nucleotide, the code is 1, then the DNA nucleotide after synthesis will be among the pyrimidine set (C, T) and if the code is 0, then the resultant DNA nucleotide will be among the purine set (A, G).

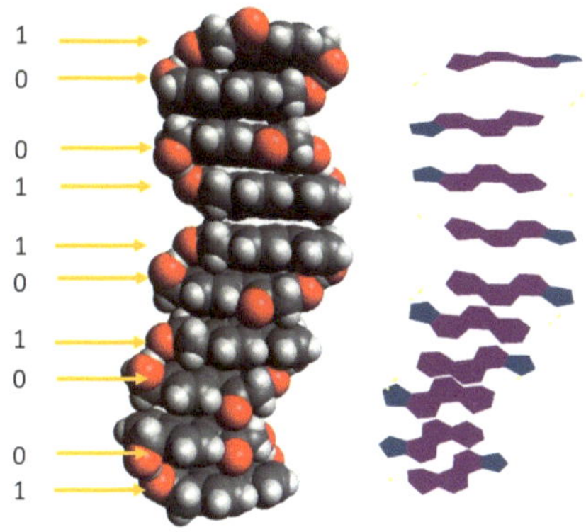

Figure A.24: Along the edge, another code is available reading from the bottom as 1001011001 where '1' represents the six-carbon ring, and '0' represents the five-carbon ring along one of the strands. Thus, the six-carbon ring will be a pyrimidine, and the five-carbon ring will be a purine, after synthesis.

These two binary codes can used to determine the identity of the five nucleotides and the identity of the corresponding five planar steroid molecules after the synthesis steps are conducted. In this case, the nucleotides are specified to be molecules 1, 3, 5, 7 and 9, and the corresponding linking steroid molecules for a circular representation will be 10, 2, 4, 6, and 8. Looking at the intermolecular match for hydrogen bonding the code for the first nucleotide comparing 10 and 1 of the code 0110100101, is a“'1” and a “0”, thus there is no match, and the molecule one after synthesis will be either (C,G). From the second code 1001011001, for the first molecule it is 1, thus it will be, after synthesis, a pyrimidine, and hence (C, T). The intersection of the two sets for molecule 1 is thus C, and after synthesis it will be Cytosine and molecule 10 will be testosterone-like. Following a similar procedure for Molecules 3, 5, 7 and 9, the result will be A, G, T, G, respectively, and Molecules 2, 4, 6, and 8 will be cortisol-like, testosterone-like, cortisol-like, testosterone-like, respectively.

A.4 Scaffolds for Encoded Structure Formation

The planar steroid assemblies and out-of-phase binding patterns provide an array of structures to associate amino acids, and with the ability to reverse encode the amino acids into DNA via planar steroid arrays enables a variety of structures. In Figure A.25(a), when the orientation of the end-groups are capable of hydrogen bonding, a zig-zag orientation results. In Figure A.25(b), alternatively, when the functional groups that are not connected to the phosphodiester-sugar linkage are hydroxyl groups, there will be an equivalent separation.

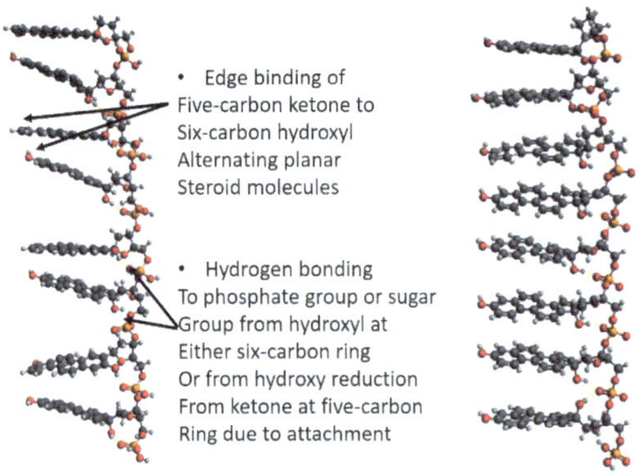

(a) (b)

Figure A.25: The planar steroid assembled from an out-of-phase binding and released produces a single-sided connected strand that is capable of self-organization, which results in templates for the fabrication of proteins in a self-catalyzed constructs, depending upon the orientation of the functional groups. (a) When the orientation of the end-groups is positioned such that hydrogen bonding is feasible, a diagonal orientation results. (b) When the functional groups that are not connected to the phosphodiester-sugar linkage are hydroxyl groups, there will be an equivalent separation.

In Figure A.26, the planar steroid templates are arranged to induce angled structures as the phosphodiester binding occurs at the six-carbon ring position adjacent to the hydroxyl group. As indicated in Figure A.26(a), the angled template configuration results because the ketone group at the five-carbon ring are positioned through a rotation process to be spatially be as far away as possible, which results in a staggered position of the ketone groups at the end, in the D-ring. The angle of the five-carbon ring contributes to this formation. Furthermore, as shown in Figure A.26(b), the connection via a hydrogen bonding mechanism from the planar steroid molecule to the sugar molecule also assists in stabilizing the orientation of the sheets at the angle where the ketone groups are positioned apart.

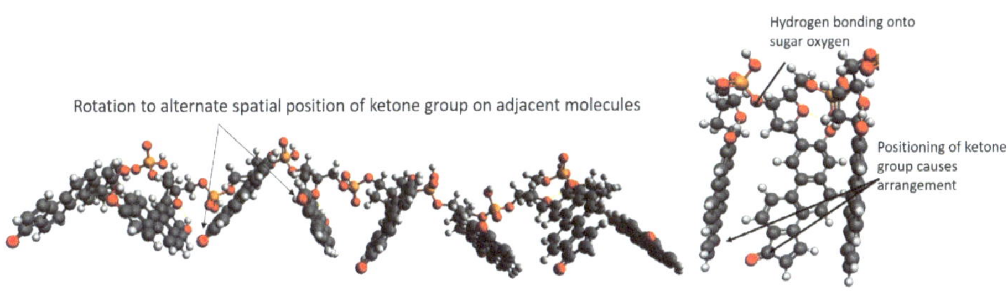

(a) Side View (b) Top, Angle View

Figure A.26: For planar steroid templates arranged to induce the formation of titled structures, the phosphodiester binding occurs at the six-carbon ring position adjacent to the hydroxyl group. (a) A diagonal template results because the ketone group at the five-carbon ring are positioned to spatially be as far away as possible, which results in a staggered position, as the planar steroid molecules have an angle form. (b) In addition, the connection of hydrogen bonding from the planar steroid molecule to the sugar molecule stabilizes the orientation of the sheets.

These are but two types of scaffolds for interaction. Basically, all components within the cell that are compatible with amino acids are candidates for scaffolds to develop encoded structures which will conform to the structure and chemical aspects of the component. In this manner, the structure is conserved and evaluated for possible future use through a genetic path.

A.5 Feedback and Feedforward Control

In addition to providing the mechanisms for generating proteins from amino acids, the planar steroid molecule has an ability to participate in a feedback loop to promote those amino acid sequences which are more likely to be of use in the surrounding environment, and thus induce increases in the availability of the molecule to further participate in molecular interactions. In Figure A.27, the feedback loop is presented, in which the planar steroid molecule binding onto the strands induces strand separation, thereby accommodating transcription of amino acids into proteins. If the proteins are utilized, it will exit the strand, and thus the planar steroid structures are available for further binding. This feedback will continue until the molecular structures maintain a capability to achieve strand separation, as the forces from repeated implementation of the amino acid groupings into the interaction vessel on the upstream strands may cause the degradation of planar steroids bound to the promoter sites, thereby terminating the strand separation, transcription and translation. Therefore, formed proteins that feedback to maintain or inhibit strand separation will also have an impact on the transcription rates.

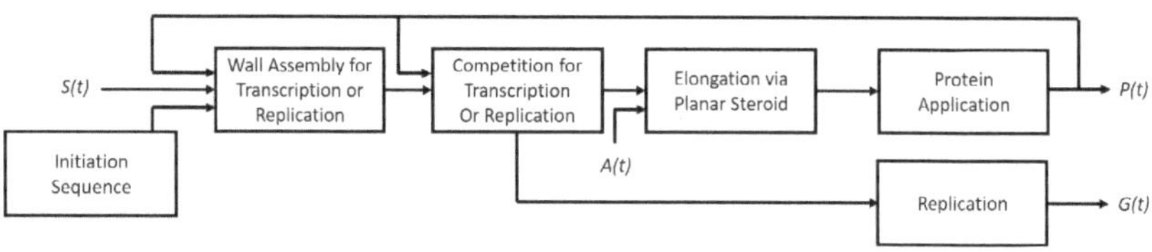

Figure A.27: The feedback loop to increase the rate of protein generation for those proteins that are more successful, that is finding greater usage. In this block diagram, the planar steroid molecule concentration triggers the strand separation, which is also a function of the enhancer and inhibitor concentration which may influence the trigger zone. $S(t)$ denotes planar steroid structures capable of inducing transcription, $A(t)$ denotes amino acid concentration over time, and $P(t)$ denotes proteins that are synthesized and diffuse away from the DNA molecule for usage in the surrounding environment. There is competition for the transcription and replication process to determine whether the protein serves a viable purpose and should proceed to replication of the genetic material $G(t)$ or be excised out of the strand.

The inherent function of the cell to perform a useful purpose and to replicate that purpose, that is to be fruitful and multiply, I refer to as cellular will. Thus, in establishing an origin of DNA and life, it is necessary to establish how cellular will is designed into the cell, and its capability to integrate within the full system.

For example. in this case, the design is described that enables a planar steroid molecule set capable of transiting to a self-organized complex system capable of a comprehensive objective, which communicates within its environment.

To develop the functions responsible for cellular will, consistent with the systems framework and how this works within the context of planar steroid assemblies and binding patterns, the two controlling factors are transcription and replication. Transcription involves the formation of protein structures, and if such structures fit within the context of the cellular purpose, replication proceeds. Within the construct of the planar steroid comprising the walls of the interaction vessel, which thereby dictates the transcription and translation properties, and in that the walls are involved in replication, it creates the situation whereby transcription and replication are in competition. If the transcription process is too dominant, then there is no time for replication, and the cell will not thrive. On the other hand, if transcription is not significant, then the cell performs no useful purpose, and although it will replicate, results in no contribution to the overall system, which also will not thrive. Thus, there is an inherent balance to the system, since the transcription and replication portions are controlled via the same process involving the same molecular set of planar steroid molecules involved in wall assembly. In Figure A.28, this concept of cellular will is presented within the context of a feedback and feedforward control mechanism.

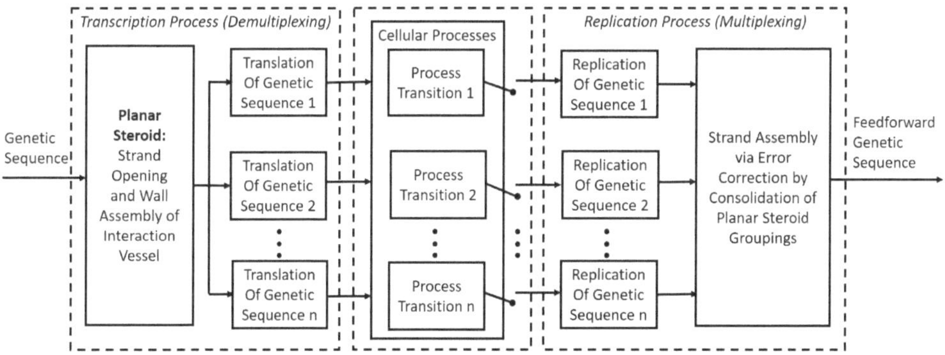

Figure A.28: The initial convergence routine to achieve a purposeful objective in which the transcription and translation processes evaluate the implementation usage, and as replication is in competition with the transcription process, there is an optimization procedure to determine if the genetic candidate is spliced out of the genetic sequence or is permitted to move forward.

As mediated through the planar steroid molecule, which functions to initiate transcription and to form the interaction vessel, including walls and ceiling, to promote translation enabling peptide bond formation, its regulation becomes a point at which cellular and later systems integration can take place. Control of the concentration and transportation of planar steroid molecules into the cytoplasm are thus able to differentiate the cellular objective, and therefore is itself a building block for further development ultimately resultant in systems performance. Consequently, a unified description of systems performance at the cellular, nuclear, and systems level is achievable through the planar steroid molecule as a central regulatory element, connecting a complex system and achieving cellular will and intercellular communication.

A.6 Direct Intermolecular Binding of Cortisol and Thymine

Calculations through molecular modeling is presented to indicate the feasibility of binding a hydroxyl group of cortisol onto the ketone group at the carbon two position of thymine, thereby completing three hydrogen bonds through an intermolecular means. Thus, AT-cortisol will match that of CG-testosterone, with each having capacity for three internal hydrogen bonds.

In Figure A.29(a), the oxygen groups of cortisol are numbered, including O2 which will bond onto thymine, as indicated in the molecular model of Figure A.29(b). The hydrogen bond distance between O2 and the oxygen group of thymine is 1.827 Å. The other bonding arrangements include O1 engaged through a water molecule with the phosphate group, although this group seems apparent for interacting with the protein molecule such as the glucocorticoid receptor. This is needed to the ionized charge of the phosphate group will prevent the direct interaction. The ionized form of the phosphate group is a transition from the initial PO_3OH original form of DNA. O3 is engaged into bonding onto the alpha position of hydrogen with the sugar molecule, and O5 is hydrogen bound to the phosphate group, with coupling of O4 into O5.

This result indicates the feasibility of binding cortisol directly onto the AT nucleotides. It requires that the AT dinucleotide rotating out of its slotted configuration to allow access for binding of cortisol. This rotational process stands to reason that the transcription factor aids in this process of distorting the chain to allow access to cortisol via the glucocorticoid receptor. This approach differs from the consensus theory of transcription, which proposes allosteric interaction. However, the results presented in this manuscript suggest that the transcription process involves direct intermolecular interaction with ligands such as the steroid hormones.

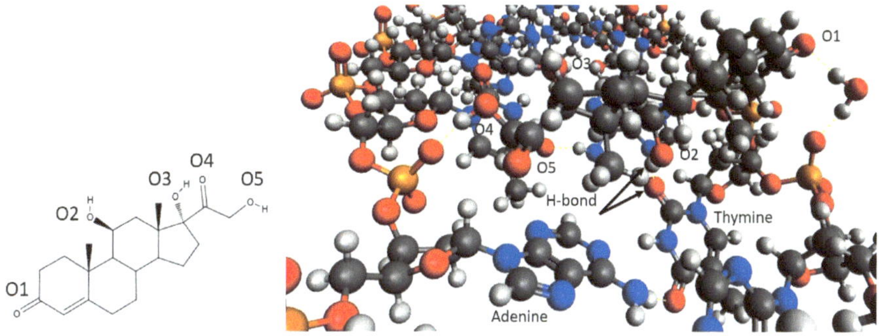

(a) (b)

Figure A.29: (a) Numbering of oxygen groups of cortisol; (b) Binding of cortisol onto Adenine-Thymine showing the position of the five oxygen groups and the intermolecular hydrogen bond of cortisol onto AT.

A.7 Color Key

Color keys for the work on the encoded sequences are also quite helpful in understanding the various structures as in effect it gives an inventory of the product composition during the steps. For example, the color key of the unified complex is indicated in Figure A.30, which shows the map to product DNA starting from the planar steroid molecule. After assembly into a unified complex, and synthesis into dinucleotides, the composition is given. As can be seen it is comprised of the dinucleotides both inverted and in the standard direction. The planar steroid molecules complete the construction. After rotation of the unoriented dinucleotides into standard position, there are two constructs: one is the planar steroid molecules comprised of either A-T complementary or C-G complementary, and the double helix DNA.

This construction associated with the DNA double helix compares against that of the bilayer cell membrane. In the color key of Figure A.31, it indicates that the planar steroid molecule is incorporated into a complex, but because the binding is out-of-phase on either strand, there is a pull-out, and thus the complexity seen in the unified complex was not reached. Further processing transforms the planar steroid into a phospholipid bilayer.

The amino acids color key is indicated in Figure A.32. More analysis is needed, but the preliminary results described in the revision indicate that the amino acids are formed through out-of-phase binding, and partial synthesis, prior to pull-out of the strands. Further degradation of the materials after pull-out may produce a full

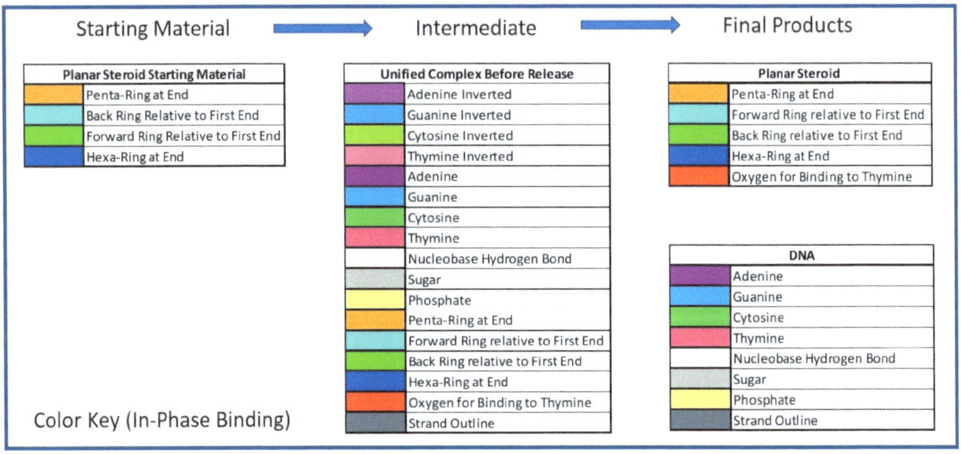

Figure A.30: The color key associated with in-phase binding of the phosphodiester chain to the unified complex. Note the complexity of the unified complex which is relieved through the transition to the DNA double helix and planar steroid molecules.

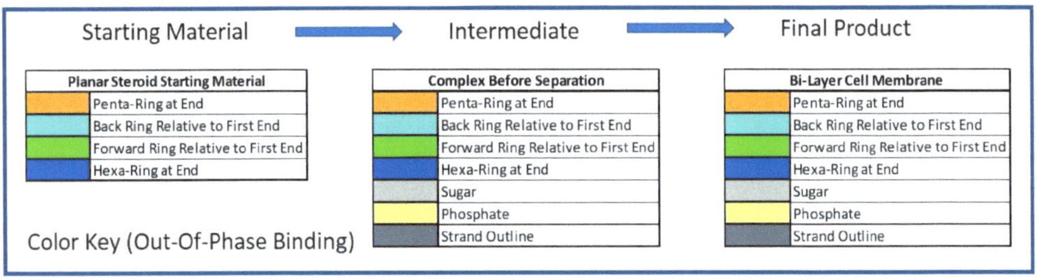

Figure A.31: he color key for the formation of the phospholipid layer of the cell membrane. Starting from the assembly of planar steroids, an out-of-phase binding of the phosphodiester linkage induces separation, and stabilization at the mid-point due to complementary binding of the ends of the planar steroid molecules.

set of amino acids, which thus will be available for integration with the interaction vessels ultimately for peptide assembly.

In Figure A.33, the color key for RNA products is presented, which includes mRNA. The color code changes for thymine of DNA to uracil of RNA, although the color code for the sugar of DNA is the same as that of the RNA even though there is hydroxy group for RNA. Consequently, the color code for sugar in mRNA products

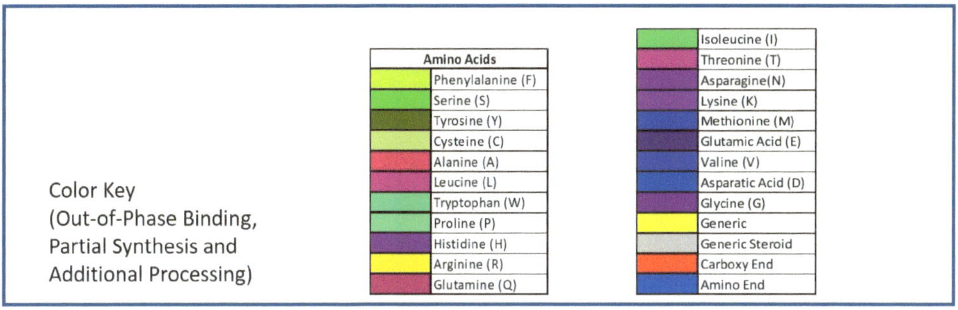

Figure A.32: Color key for amino acids, which may be formed from an out-of-phase binding to the assembly of planar steroids, followed by partial synthesis, separation, and further reduction of the bond structure.

needs to be taken in context of the description. The discussion on tRNA and rRNA involve the constructs taken from DNA structures, and thus may contain a thymine structure.

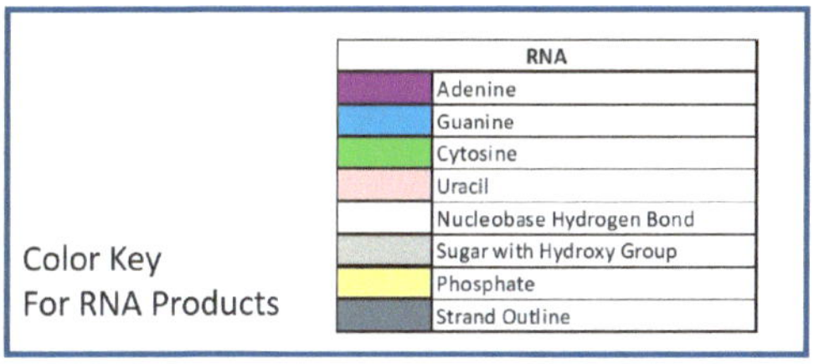

Figure A.33: Color key for RNA products. The color code changes for thymine of DNA to uracil of RNA. The color code for the sugar of DNA is the same as that of the RNA even though there is hydroxy group for RNA; thus, the color code for sugar needs to be taken in context of the description.

Appendix B

Methods

- Molecular modeling software: The software program Avogadro was utilized to build the three-dimensional molecular models, position the elements, and calculate the bond distances. An optimization routine based upon steepest descent was implemented to automatically position the elements and the interactive molecular configurations. The website Molview.org was used for several chemical wireframe drawings and the Wikipedia page on DNA was used as an initial reference for the ACTG wireframe used in the overlays. Matlab was used to generate the energy dissipation requirements when the DNA nucleotides were connected by the planar steroid molecule. Matlab was also used to produce a random number sequence to initiate the results used in the example of twenty-five dimers of nucleotides aligned with planar steroid molecules comprising a unified complex.

- Development of molecular models: The DNA nucleotides and steroid molecules were developed and synthesized cooperatively in an intermolecular manner by stacking dimers with the internal two rings aligned, and the end rings arranged via hydrogen bonding. The intermolecular hydrogen bonds were configured, and then the optimization software was used to orient the molecules, positioning the dimers to minimize energy. An assembly was then produced of stacking dimers on top of each other. After establishing the stack, the phosphodiester backbone was assembled starting from the bottom at molecule 1, and then linking 3, 5, 7 and 9 by forming the N-C-O coupling, which will ultimately correspond to the nucleotide to sugar link. These results were performed in conjunction with the synthesis scheme described in the Appendix. After coupling, the sugar molecule was formed. Both sides were completed, with hydrogen bonding stabilizing the planar steroid molecule to the phosphodiester structure by hydrogen bonding. The internal nitrogen elements of the purine and pyrimidine components of the nucleotide pairs were then positioned accordingly, and the bonds were broken with the addition of oxygen groups selected at the appropriate position on thymine, cytosine, adenine, and guanine. The rotation was performed on the nucleotide pairs according to the codes thereby ejecting the co-synthesized intercalated planar steroid molecules. Transcriptional positioning was performed by inserting the resultant steroid into the appropriate slot and calculating the energy reduction from the stabilization. For the ionic study, Mg^{2+} ions were included to stabilize the ends to the phosphodiester chain.

- Experimental: To obtain the relative activity of the synthetic steroids and steroid hormones, the literature was surveyed, and representative values were selected for the activity of prednisone, prednisolone, cortisol, and dexamethasone [30]. The experimental structures of these steroid hormones were then overlaid to the experimental structures of the A-T nucleotide pair to achieve a third hydrogen bond by intermolecular coupling to the corticosteroid.

- Overall Procedure: The development of this approach took the path of a top-down investigation, getting to the bottom, and then taking a bottoms-up approach. The initiating point was a mathematical analysis of a coordination of the nervous and cardiovascular systems to characterize a fever induced by pneumococcal vaccination in conjunction with exercise [31]. An investigation into the chemistry of the interaction resultant in the fever led to the concept of inhibition of cortisol by prostaglandins, which also led to the concept of coupling ionic means at the glucocorticoid receptor to explain the differences in activity between the two molecules [32]. In further pursuit of a fundamental understanding of a fever resulted in transporting the glucocorticoid receptor to the nucleus to evaluate possible transcription of cortisol and prostaglandins through binding directly onto the DNA nucleotides. An evaluation of the structural symmetry of cortisol and the adenine-thymine base pair was determined, noting the capacity for intermolecular binding, which would thereby explain noted differences in activity. Thus the symmetry of steroid molecules and DNA nucleotides was discovered [17, 18, 19]. This structural symmetry of steroid molecules and DNA nucleotides then led to the current results to define a basis for the transcription results perceived of cortisol on adenine-thymine, and other potential interactions between RNA nucleotides and other functional cyclic compounds. Hence, the original mathematical basis describing the systems and cellular responses are consistent with the nuclear developments, and thus enable an integrated solution. A bottoms-up analysis then was performed to determine the genetic code, and the other binding patterns associated with the phospholipid bilayer and other biological materials. This analysis procedure, which started from as a top-down approach, now having made its way to DNA analysis, and its origin, is continuing in a bottoms-up manner to eventually reach the starting point, and hence will result in an fully integrated systems solution.

Appendix C

Encoding Sequence for DNA - Steroid Complex in Example

In Figure C.1, the encoded sequence that was used in the example of Chapter 7 is presented. It is comprised of a set of 25 pairs, hence a total of 50 planar steroid molecules. This encoded set was specified by a third party, which was a random number generator with uniform distribution. After processing, it results in the nucleotide sequence GAAATAGCACTGTCTCGTCGGTTGG with the first molecule of the sequence considered as the first nucleotide. For the first molecule since the preceding planar steroid structure was not specified, it was considered that the last planar steroid molecule would determine if hydrogen bonding would be present in the first planar steroid molecule, as it was destined to be a nucleotide.

It is noted that this sequence can be produced easily if the nucleotide sequence had been given instead. It would start from a G arrangement, which would be either PFBK or PBFK depending on the sequence of a preceding planar steroid molecule, which for this problem was considered as the fiftieth molecule, KFBP; thus the first molecule would be PFBK, since it would not have hydrogen bonding, and thus there is only one choice from Table 3.2. This corresponds to the planar steroid orientation KFBP. Thus, to achieve the next nucleotide, A, would involve selecting the appropriate sequence using Table 3.2. It requires hydrogen bonding since A-T only has two internal hydrogen bonds and three are required, and A is a pyrimidine, thus it must start with P. Thus, the choice is PBFK. This process is repeated until the sequence is finished.

P	F	B	K
K	F	B	P
P	B	F	K
K	B	F	P
P	F	B	K
K	F	B	P
P	B	F	K
K	B	F	P
K	B	F	P
P	B	F	K
P	B	F	K
K	B	F	P
P	B	F	K
K	B	F	P
K	F	B	P
P	F	B	K
P	F	B	K
K	F	B	P
K	B	F	P
P	B	F	K
K	F	B	P
P	F	B	K
P	B	F	K
K	B	F	P
K	B	F	P
P	B	F	K
K	B	F	P
P	B	F	K
K	F	B	P
P	F	B	K
K	F	B	P
P	F	B	K
P	B	F	K
K	B	F	P
K	B	F	P
P	B	F	K
K	B	F	P
P	B	F	K
P	F	B	K
K	F	B	P
P	F	B	K
K	F	B	P
K	F	B	P
P	F	B	K
K	B	F	P
P	B	F	K
P	F	B	K
K	F	B	P
P	F	B	K
K	F	B	P

Figure C.1: The encoding sequence for the example of Chapter 7 on encoding, transmitting, decoding and translating. To take a neutral example, the sequence of dimers was generated from a third party, that is a uniformly distributed random number generator repeated 25 times to produce an encoded sequence of the ring orientation of the dimers, thus resulting in a unified complex of 50

Appendix D

Statement of No Conflicts of Interest

The author asserts no conflicts of interest. No external funding was used for this study.

Bibliography

[1] F. Dyson, *Origins of life.* Cambridge University Press, 1999.

[2] A. Lazcano and S. L. Miller, "The origin and early evolution of life: prebiotic chemistry, the pre-rna world, and time," *Cell*, vol. 85, no. 6, pp. 793–798, 1996.

[3] J. D. Sutherland, "Opinion: Studies on the origin of life—the end of the beginning," *Nature Reviews Chemistry*, vol. 1, no. 2, pp. 1–7, 2017.

[4] R. Cojocaru and P. J. Unrau, "Origin of life: Transitioning to dna genomes in an rna world," *eLife*, vol. 6, p. e32330, 2017.

[5] P. Forterre, J. Filee, and H. Myllykallio, "Origin and evolution of dna and dna replication machineries," *Madame Curie Bioscience Database [Internet]. Landes Bioscience; 2000-2013.*, 2013, https://www.ncbi.nlm.nih.gov/books/NBK6360/.

[6] W. Gilbert, "Origin of life: the rna world," *Nature*, vol. 319, no. 20, p. 618, 1986.

[7] H. M. Goodman and A. Rich, "Formation of a dna-soluble rna hybrid and its relation to the origin, evolution, and degeneracy of soluble rna," *Proceedings of the National Academy of Sciences of the United States of America*, vol. 48, no. 12, p. 2101, 1962.

[8] S. L. Miller and J. L. Bada, "Submarine hot springs and the origin of life," *Nature*, vol. 334, no. 6183, pp. 609–611, 1988.

[9] S. Nakashima, Y. Kebukawa, N. Kitadai, M. Igisu, and N. Matsuoka, "Geochemistry and the origin of life: From extraterrestrial processes, chemical evolution on earth, fossilized life's records, to natures of the extant life," *Life*, vol. 8, no. 4, p. 39, 2018.

[10] K. A. Lanier and L. D. Williams, "The origin of life: models and data," *Journal of molecular evolution*, vol. 84, no. 2-3, pp. 85–92, 2017.

[11] S. Hoshika, N. A. Leal, M.-J. Kim, M.-S. Kim, N. B. Karalkar, H.-J. Kim, A. M. Bates, N. E. Watkins, H. A. SantaLucia, A. J. Meyer *et al.*, "Hachimoji dna and rna: A genetic system with eight building blocks," *Science*, vol. 363, no. 6429, pp. 884–887, 2019.

[12] R. Service, "Researchers may have solved origin-of-life conundrum," *Science*, 2015, doi:10.1126/science.aab0325.

[13] B. H. Patel, C. Percivalle, D. J. Ritson, C. D. Duffy, and J. D. Sutherland, "Common origins of rna, protein and lipid precursors in a cyanosulfidic protometabolism," *Nature chemistry*, vol. 7, no. 4, p. 301, 2015.

[14] Á. Kun and Á. Radványi, "The evolution of the genetic code: Impasses and challenges," *Biosystems*, vol. 164, pp. 217–225, 2018.

[15] E. V. Koonin and A. S. Novozhilov, "Origin and evolution of the universal genetic code," *Annual review of genetics*, vol. 51, pp. 45–62, 2017.

[16] P. Marshall, *Evolution 2.0: Breaking the Deadlock Between Darwin and Design.* BenBella Books, Inc., 2015.

[17] C. Schaper, "Endogenous Binding of Steroid Molecules to DNA Nucleotides by a Ca2+/PO4- Process to Enable Gene Transcription," *ChemRxiv*, 2020, https://doi.org/10.26434/chemrxiv.11868261.v2.

[18] ——, "Structural Symmetry of DNA Nucleotides and Steroid Hormones," *ChemRxiv*, 2020, https://doi.org/10.26434/chemrxiv.11991567.v1.

[19] ——, "New Code for DNA Nucleotide Sequences," *ChemRxiv*, 2020, https://doi.org/10.26434/chemrxiv.12003786.v1.

[20] D. Becker, "Basic and clinical pharmacology of glucocorticosteroids," *Anesthesia Progress*, vol. 60, no. 1, pp. 25–32, 2013.

[21] C. Schaper, "Intermolecular Binding of Functional Cyclic Compounds to RNA Nucleotides by an Ionic Process," *Preprints*, 2020, https://doi.org/10.20944/preprints202004.0522.v1.

[22] N. Hoffmann, "Photochemical reactions as key steps in organic synthesis," *Chemical Reviews*, vol. 108, no. 3, pp. 1052–1103, 2008.

[23] U. Reuther and A. Hirsch, "Synthesis, properties and chemistry of aza [60] fullerene," *Carbon*, vol. 38, no. 11-12, pp. 1539–1549, 2000.

[24] O. Vostrowsky and A. Hirsch, "Heterofullerenes," *Chemical reviews*, vol. 106, no. 12, pp. 5191–5207, 2006.

[25] Y. Li, G. Zhang, D. Wang, B. Xu, D. Xu, N. Lou, and L. Gan, "Fullerene-based macro-heterocycle prepared through selective incorporation of three n and two o atoms into c60," *Angewandte Chemie International Edition*, vol. 55, no. 47, pp. 14 590–14 594, 2016.

[26] J. B. Siegel, A. Zanghellini, H. M. Lovick, G. Kiss, A. R. Lambert, J. L. S. Clair, J. L. Gallaher, D. Hilvert, M. H. Gelb, B. L. Stoddard *et al.*, "Computational design of an enzyme catalyst for a stereoselective bimolecular diels-alder reaction," *Science*, vol. 329, no. 5989, pp. 309–313, 2010.

[27] A. Sanyal and J. K. Snyder, "Stereoselective diels- alder reactions of chiral anthracenes," *Organic letters*, vol. 2, no. 16, pp. 2527–2530, 2000.

[28] S. E. Wheeler, A. J. McNeil, P. Muller, T. M. Swager, and K. Houk, "Probing substituent effects in aryl- aryl interactions using stereoselective diels- alder cycloadditions," *Journal of the American Chemical Society*, vol. 132, no. 10, pp. 3304–3311, 2010.

[29] K.-C. Chou, "Low-frequency vibrations of dna molecules," *Biochemical Journal*, vol. 221, no. 1, pp. 27–31, 1984.

[30] J. Labrin and K. Cunningham, "What corticosteroid is most appropriate for treating acute exacerbations of copd," *The Hospitalist*, June 2011.

[31] C. Schaper, "Dynamics and Internal Control of Body Temperature in Response to Infectious Agents and Other Causal Factors," *bioRxiv*, 2019, https://doi.org/10.1101/566679v1.

[32] ——, "Competitive Inhibition of Cortisol by Prostaglandins at the Ligand Binding Domain of Glucocorticoid Receptors," *bioRxiv*, 2019, https://doi.org/10.1101/851501.

www.ingramcontent.com/pod-product-compliance
Lightning Source LLC
LaVergne TN
LVHW072348011026
849160LV00021B/21
9781735372167